QUANTUM PHYSICS FOR BEGINNERS

The Easy Guide to Understand how Everything Works through the Behavior of Matter, the Law of Attraction and the Theory of Relativity.

Author: Ethan Hayes

© **Copyright 2020 Ethan Hayes - All rights reserved.**

The content contained within this book may not be reproduced, duplicated or transmitted without direct written permission from the author or the publisher.

Under no circumstances will any blame or legal responsibility be held against the publisher, or author, for any damages, reparation, or monetary loss due to the information contained within this book. Either directly or indirectly.

Legal Notice: This book is copyright protected. This book is only for personal use. You cannot amend, distribute, sell, use, quote or paraphrase any part, or the content within this book, without the consent of the author or publisher.

Disclaimer Notice: Please note the information contained within this document is for educational and entertainment purposes only. All effort has been executed to present accurate, up to date, and reliable, complete information. No warranties of any kind are declared or implied. Readers acknowledge that the author is not engaging in the rendering of legal, financial, medical or professional advice. The content within this book has been derived from various sources. Please consult a licensed professional before attempting any techniques outlined in this book.

By reading this document, the reader agrees that under no circumstances is the author responsible for any losses, direct or indirect, which are incurred as a result of the use of information contained within this document, including, but not limited to, errors, omissions, or inaccuracies.

Published in the United States of America

Table of Contents

INTRODUCTION _____ 1
CHAPTER 1: WHAT IS QUANTUM PHYSICS AND WHY I SHOULD LEARN IT _____ 11
CHAPTER 2: INTRODUCTION TO THE MAIN ELEMENTS _____ 22
CHAPTER 3: HEISENBERG LIMITED STATISTICS _____ 32
CHAPTER 4: PARTICLE ENTANGLEMENT _____ 37
CHAPTER 5: ANGULAR MOMENTUM _____ 41
CHAPTER 6: BACK TO BASICS _____ 46
CHAPTER 7: IS IT A WAVE OR A PARTICLE _____ 58
CHAPTER 8: HOW LIGHT BEHAVES _____ 81
CHAPTER 9: THE THEORY OF RELATIVITY _____ 90
CHAPTER 10: QUANTUM THEORY _____ 96
CHAPTER 11: ELASTIC SHOCKS AND INELASTIC SHOCKS _____ 113
CHAPTER 12: PRACTICAL APPLICATIONS _____ 125
CHAPTER 13: ENERGY _____ 141
CHAPTER 14: PHILOSOPHICAL IMPLICATIONS _____ 150
CHAPTER 15: LAW OF ATTRACTION _____ 158
CHAPTER 16: QUANTUM PHYSICS AND HEALTH _____ 165
CHAPTER 17: STRING THEORY AND THE THEORY OF EVERYTHING _____ 171
CHAPTER 18: M-THEORY _____ 175
CHAPTER 19: BLACK HOLES _____ 185
CONCLUSION _____ 190

Introduction

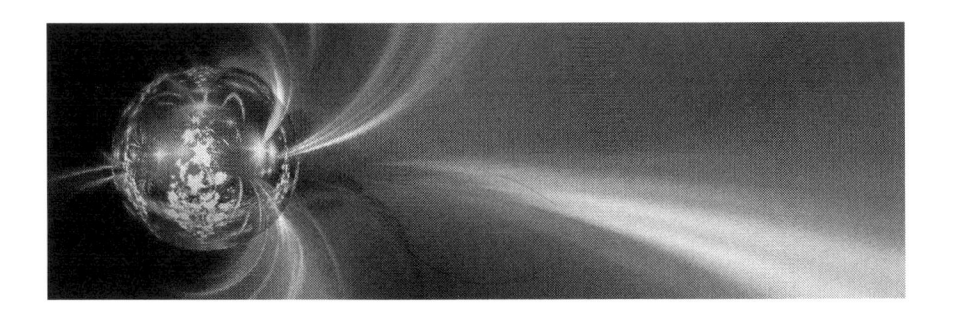

No matter who you are and what you do for a living, quantum physics will bring a whole new perspective into your life on so many things that it is impossible to ignore it. Quantum physics is the underlying framework for the laws of nature that explains matter and energy's behavior on the subatomic level. We as a whole move to the quantum tune, and all work in the same way. If you need to clarify how electrons travel through a PC chip, how light photons in a sunlight-based board are changed over into power or intensified into a laser, or how the sun keeps on consuming, you need to use quantum physics.

Quantum physics confirms that a thing can only exist if it is observed. The quantum is organized according to the principles that influence the observer's mind. In this context, it's important to define quantum in general terms. It is a whole set of knowledge that considers the smallest particles in nature. Their behavior and condition allow them to interact with each other, resulting in one of the fundamental concepts: quantization. It is generally understood as the departure from a set of concrete values to attain discrete ones. Quantum is more accurately and concisely detailed by the end of the first chapter.

When something is observed, quantum merges into subatomic particles. Then into atoms, followed by molecules. Until finally, something in the physical world manifests itself as a localized temporal spacetime experience that can be perceived through the mediation of our five physical senses. It would then lead to something that appears to be reliable and is part of what people usually understand as physical reality. As energy, every single thought directly and instantly influences the quantum field, whereby Quantum merges into a localized, observable experience event, object, or other influence.

This process is the basis for how everyone creates their reality. Those who understand and comply with universal laws are conscious creators, while others create their life experience by default. As a result, they attribute everything experienced as a consequence of their unconscious thinking to superstitious beliefs such as luck, fate, chance, and fortune. We know that conscious creation is also the basis of the law of attraction and the law of cause and effect. We will talk more about this in Chapter 15.

Quantum Physics is more about explaining the microscopic world with equations to match observations, then observations matching equations. As someone who finds this microscopic world and all the parts of it fascinating, it quickly becomes apparent that your math skills need to be sharpened to understand this area of study. Not only that, the discoveries made at the turn of the 20th century that kicked off the study of quantum physics have had enormous repercussions on philosophy, our way to interpret reality, but even literature, music, and last but not least, sci-fi, comics, and movies.

Whether you are looking for a general overview of the subject or an introduction to one of the most fascinating subjects the human mind has ever thought of, this book is the perfect choice for you. I want to share with you all the amazing discoveries of quantum physics and their philosophical and technological applications and, I promise you, you won't be disappointed. When you put down this book, you won't only have a basic but precise knowledge of quantum physics, but you'll find yourself looking at the world with new eyes.

While it is true that for a complete and profound knowledge of quantum physics, a study of mathematics is necessary, you can gaze and marvel at the wonders of the universe, even with a basic or non-existent knowledge of mathematics. Actually, the most interesting concept of quantum physics is philosophical, and to understand them, you will only need your curiosity and thirst for knowledge.

This book aims to make quantum physics a fun and rewarding experience for anybody. It is also meant as an introduction to the most compelling physics dilemmas of our times. Hopefully, it will shed some light and give you some insight into this fascinating subject that has been treated as a matter for geniuses and math wizards for too long.

Quantum physics is a realm that seems unreachable and beyond understanding to many people, but it's actually an extremely fascinating branch of science. A lot of quantum physics is still largely undiscovered or unexplained, and that's what makes it so charming!

Throughout this book, we'll explore quantum experiments and theories, both how they came to be and then how they have grown to become critical parts of what we now know as quantum physics.

This book will give a concise and clear exposition of quantum physics. Keep an open mind, and you will find a whole new world. Let's dive in and uncover the basics of quantum physics!

Key concepts discussed in this book

Throughout this book, you will discover many exciting theories and get introduced to the great minds behind them. I intend to make all this easy to understand for anyone. While at first, some of these concepts and theories might seem too hard to comprehend, as you proceed through the book and refine your understanding, they will start making more sense. I want to start this book with a brief introduction to some of the key concepts and names, as this knowledge will help you navigate through the book and get the most out of it.

The Atom

We take atoms for granted. Their existence was still controversial until the beginning of the 20th century. Already in the 5th century BC, the ancient Greeks, especially Leukipp and his pupil Democritus, spoke of atoms. They thought the matter was made up of tiny, indivisible units. They called these atoms (ancient Greek "átomos" = indivisible).

It wasn't until the end of the 18th century that science provided concrete evidence of atoms' existence. In 1789 Antoine Lavoisier formulated the law of conservation of mass. 10 years later Joseph Louis Proust formulated the law of definite proportion, until John Dalton built upon these theories to develop the law of multiple proportions. His publication in 1805 marked the beginning of the scientific atomic theory.

In his miracle year 1905, Albert Einstein not only presented the special theory of relativity and solved the mystery of the photoelectric effect, but he was also able to explain the Brownian motion. In 1827 the Scottish botanist and physician Robert Brown (1773 - 1858) discovered that dust particles only visible under the microscope make jerky movements in the water. Einstein was able to explain this by the fact that much smaller particles, which are not visible even under the microscope, collide in huge numbers with the dust particles and that this is subject to random fluctuations. The latter leads to jerky movements. The invisible particles must be molecules. Therefore, the Brownian movement's explanation was regarded as their validation and thus also as the atoms' validation.

In chapter 2 and 6 we will look more closely at the Rutherford, Thomson and Bohr's models and how they contributed to the formulation of the Quantum Atomic Theory.

Copenhagen Interpretation

Physicist Niels Bohr first proposed the Copenhagen interpretation in 1920. In very simple terms, it says that a quantum particle doesn't exist in one state or another, but in all of its possible states at once. To this day, many quantum physicists still assume the Copenhagen interpretation is correct.

Planck's Radiation Law and Planck Constant

Planck's radiation law is a mathematical equation proposed by the German philosopher Max Planck to describe the distribution of energy that reflects the energies of black body radiation (the body of thought reaches a certain degree of measurement and regains energy as it absorbs).

The Planck Constant (referred to as h) was introduced in 1900 and it is a fundamental physical constant characteristic of the mathematical formulations of quantum mechanics. It describes particles and waves' behavior on the atomic scale, including the particle aspect of light. The significance of Planck's constant is that radiation, such as light, is emitted, transmitted, and absorbed in discrete energy packets, or quanta, determined by the frequency of the radiation and the value of Planck's constant. We will refer to this constant many times over the next chapters.

De Broglie hypothesis

Before we come to the wave mechanics of Erwin Schrödinger, we have to talk about the French physicist Louis de Broglie (1892 - 1987). In his doctoral thesis, which he completed in 1924, he made a bold proposal. Until then, wave-particle dualism was a characteristic exclusively of electromagnetic radiation.

Why, so de Broglie, shouldn't it also apply to matter? So why should matter not also have a wave character in addition to its real particle character? The examination board at the famous Sorbonne University in Paris was unsure whether it could approve it and asked Einstein.

He was deeply impressed that de Broglie got his doctorate. However, he could not present any elaborated theory for the matter waves.

It becomes a puzzle as we try to understand if all matters also exhibit the Wave-like properties. De Broglie's hypothesis was an attempt to settle this and it suggests that all matter exhibits wave-like properties and gives a relation between wavelength of matter and momentum.

After Albert Einstein made the proposition of the photon theory, physicists tried to understand if this theory was only related to light or if other material objects could still be able to display this wave-like property.

The Schrödinger Equation

The equation was coined from the name of Erwin Schrödinger. He hypothesized the equation in 1925 and made it available in 1926. It is a linear partial differential equation used for describing the state function or wave effect of a quantum system. It is relevant in discovering notable landmarks obligatory for the development of that course. In Chapters 7 and 10 we will look at this equation in much greater details.

Heisenberg uncertainty principle

Heisenberg was able to develop matrix mechanics in 1925, following the works of Hendrik Kramers. This served as a replacement for the old quantum theory as it laid the foundation for modern quantum mechanics. The main idea of this was that the

classical concept of motion doesn't serve in that qua
This is because electrons in an atom do not travel on sharply defined orbits. Simply put, the uncertainty principle states that there is a fundamental limit to what one can know about a quantum system. For instance, the more precisely you know a particle's position, the less you can know about its momentum, and vice versa. We will talk again about this principle in Chapter 3 and throughout the book.

Dirac symbol

Paul Dirac, who might be the most productive 20th-century physicist after Einstein, showed that the Schrödinger model and Heisenberg's matrix mechanics are equivalent, and are simply two different languages for a good description of the same reality. Dirac expanded Schrödinger mathematics, which now takes into account extremely high electron velocities as well (Dirac equation).

For this purpose, elements of Einstein's theory of special relativity had to be introduced, which relates to very high velocities up to the highest possible speed, the speed of light. Dirac's model also mathematically substantiated the existence of spin (predicted by Pauli, more on this in Chapter 3) and the phenomenon of wave-particle duality. It also led to several important predictions, which were proved later, such as the electron's magnetic moment and the electron antiparticle's existence – the positron.

Dirac also suggested that all other particles of matter should have anti-doubles as well, thus predicting antimatter, the existence of which was subsequently confirmed.

EPR paradox

The EPR paradox, a name given to the Einstein-Podolsky-Rosen paradox, is a paradox that helps to define and show the inherent paradoxes in the early formulas of quantum mechanics. An example of this problem was the issue of quantum entanglement, a concept that meant that two particles could get tangled with each other. Each particle is not clearly defined until it is measured. The state of this particle is defined, and by default, so is the particle it is tangled with.

This paradox came about due to a debate between Einstein and Niels Bohr, as Einstein wasn't satisfied with the ideas proposed by Bohr and his contemporaries. This was a way to prove that quantum theory was not consistent with other physics laws as was accepted at that time. We will talk more about the EPR Paradox in Chapter 6 and 9 and throughout the book.

Chapter 1: What is Quantum Physics and Why I Should learn It

It is a pretty well-known fact that Albert Einstein was not a big aficionado of the quantum mechanics theories that were shaping up during his lifetime. Time proved him wrong in some ways because some of the quantum theories are actually being proven step by step. Beyond that, however, the questions posed by Einstein are still valid - and they provide quantum researchers with a point of orientation when it comes to the answers they are yet to give. If Albert Einstein was alive today, he would probably have "converted" to quantum physics - because even throughout his life, his views on this theory changed. If his theory completely clashed with quantum mechanics at first, he used quantum concepts to explain some of his theories later on in life.

More specifically, in 1935, his experiments revealed what he called "spooky action at a distance" - or, in other terms, quantum entanglement. He then continued his experiments furthering the theory that quantum entanglement was only possible in certain circumstances. Unfortunately, however, he never got a clear answer to this follow-up, and it was left to future generations to reconcile the theories. It would be more than interesting to see what he would have to say today about the more recent discoveries and quantum mechanics experiments.

Why Do We Accept Quantum Mechanics?

Without a doubt, Einstein's work reshaped the world in so many ways that it would take an entire library of books to explain them in plain English simply. In the scientific community (and, dare we say, outside of it too), Einstein is seen as a sort of demi-god, an irrefutable authority that nobody dares to touch.

If Einstein's theory of relativity is so well-regarded and accepted, why do we even bother with quantum mechanics, then? What demon sets so many contemporary scientific researchers on the path of actually trying to reconcile the worlds of classical physics and quantum physics?

The reason quantum mechanics is accepted and still a topic of discussion is that it manages to solve what classical physics couldn't. And, as it has been shown, it would manage to push the boundaries

of knowledge and technology beyond the edges of the imagin
a spectrum, we only dared to touch our thoughts until not very long ago. September 7, 2014, might have seemed like any other day of fall in the Northern Hemisphere. The leaves were probably slightly yellow by then, and the heat of the summer was slowly starting to wear out. Maybe it even rained a little in the morning, and by the time cities were waking up to life, the fog of a slightly chillier night vanished, leaving room for a perfect day of autumn.

Everyone must know that September 7, 2014, was the day the Theory of Everything officially saw the light of day. You might have heard about it because there was a movie on the life of Stephen Hawking. Or you might have even stumbled upon it long before the movie came out.

However, what is important is that the Theory of Everything is one of the most important attempts at unifying both the theory of relativity and the quantum theory. What was initiated in the 1920s by Albert Einstein started to make sense eight decades later under the hands of Stephen Hawking.

The Theory of Everything is, perhaps, one of the most ambitious projects ever. It is one of the theories that is bound to change every single little thing - not just in physics, but in science as a whole, and, soon enough, in humanity's perception of pretty much every area of their lives.

The Theory of Everything tries to finally build a bridge between quantum mechanics and the theory of relativity. Some would even dare to say that it will "tell the mind of God" (Marshall, 2010) and that it will hold the key to humanity answering the questions it has been trying to answer for a very, very long time now.

There are several candidates for the Theory of Everything. Some of them are implausible to be proven in the equation or in practice, but some of them stand out as sane options that might be the final answer to everything.

Out of these, we would like to take the time to name the two most important contenders. As we draw close to the conclusion, we believe it is important for you to know what the most important work in physics is doing now and as such, we will take the time to expand, just a little bit, on these two theories.

One of them is called "String Theory" and what it says is that there is a ten-dimensional space we are living in. That sounds more than mind-boggling, we know, but wait until you hear more of it (if you're impatient, you can jump to chapter 17, but I suggest you keep reading on). In essence, the Theory of Everything relies on quantum gravity and it aims to address a wide range of questions in fundamental physics, such as what is going on with black holes, how the universe was formed, how to improve nuclear physics, and how to handle condensed matter physics better.

Ideally, string theory will unify gravity and particle physics (one of the main points that have to be bridged between classical physics and quantum mechanics). At the moment, however, it is not clear how much of this theory can be adapted to the real world and how much of it will allow for changes in its details.

The other theory competing with string theory for the title of "The Theory of Everything" is the Loop Quantum Gravity Theory. This paradigm is heavily based on Einstein's work, and it was elaborated towards the middle of the 1980s. To understand it, you need to remember the fact that, according to Einstein, gravity is not a force per se, but a property of space-time.

Up until the Loop Quantum Gravity Theory, there have been several attempts to prove that gravity can be treated as a quantum force, like electromagnetism or the nuclear force, for example. However, these attempts have failed.

If physicists manage to prove the Loop Quantum Gravity Theory, space-time will be pictured with space and time being granular, which would consequently mean that a minimum space exists. In other words, according to the Loop Quantum Gravity Theory, space is made out of a fine fabric of woven finite loops called "spin networks."

Although String Theory seems to be a lot more popular in mainstream media (mainly because some of its proponents are quite popular themselves, even outside scientific circles - like Michio Kaku, for example), the Loop Quantum Gravity Theory should not be dismissed in any way. Most of its implications are related to the birth of the universe, the reason for which it is also called the Big Bang Theory - and, perhaps, the reason for which the eponymous TV show was called that way as well.

The Evolution of Quantum Physics

During the 19th century, crystallographers and chemists tried to prove the existence of atoms. Still, it was not until the beginning of the 20th century that they were finally brought to light, thanks to X-ray diffraction. To model them, the quantification of the matter is a compulsory passage, which gives rise to quantum physics.

Quantum physics has brought a conceptual revolution with repercussions up to philosophy (questioning determinism) and literature (science fiction). It has enabled several technological applications: nuclear energy, medical imaging by nuclear magnetic resonance, diode, transistor, integrated circuit, electron microscope, and laser. A century after its conception, it is widely used in research in theoretical chemistry (quantum chemistry), in physics (quantum mechanics, quantum field theory, condensed matter physics, nuclear physics, particle physics, quantum statistical physics, astrophysics, quantum gravity), in mathematics (formalization of field theory) and, recently, in computer science (quantum computer, quantum cryptography). Quantum physics is known to be counter-intuitive, to shock common sense, and to require arduous mathematical formalism. Feynman, one of the greatest theoreticians specializing in quantum physics in the second half of the 20th century, wrote:

"I think I can say that no one understands quantum physics."

The main reason for these difficulties is that the quantum world (limited to the infinitely small, but which can have repercussions on a larger scale) behaves very differently from the macroscopic environment to which we are accustomed. Some fundamental differences that separate these two worlds are, for example:

- Quantification: A certain number of observables, for example, the energy emitted by an atom during a transition between excited states, are quantified, that is, to say that they can only take their

value in set discrete results. Conversely, classical mechanics most often predict that these observables can take on any amount continuously.
- Wave-particle duality: The notions of wave and particle, which are separate in classical mechanics, become two facets of the same phenomenon, described mathematically by its wave function. In particular, experience proves that light can behave like particles (photons, highlighted by the photoelectric effect) or like a wave (radiation-producing interference) depending on the experimental context, electrons, and other particles can also behave in an undulatory manner.
- In particular, it is impossible to obtain high precision in measuring a particle's velocity without getting mediocre accuracy in its position, and vice versa. This uncertainty is structural and does not depend on the experimenter's care not to disturb the system; it constitutes a limit to the accuracy of any measuring instrument.
- The observation influences the observed system: During the measurement of an observable, a quantum system sees its modified state. This phenomenon, called wave packet reduction, is inherent in the size and does not depend on the experimenter's care not to disturb the system.
- Nonlocality or entanglement: Systems can be entangled so that interaction in one place in the system has immediate repercussions in other areas. This phenomenon contradicts the special relativity for which there is a speed limit for the propagation of all information, the speed of light; however, nonlocality does not allow data to be transferred.

- Contra factuality: Events that could have happened, but did not occur, influence the results of the experiment.

Between 1925 and 1927, several physicists succeeded in assembling the pieces of the giant quantum physics puzzle in a few years. First of all, Louis de Broglie and Erwin Schrödinger reason that if light, which is a wave, can exist in the form of energy packets, then conversely, perhaps particles like electrons, which are of small energy packets, can behave like a wave. They thus develop wave functions.

A particle can, therefore, exist in several superimposed states, waves, or particles. It is the principle of superposition. Werner Heisenberg introduces the principle of uncertainty, according to which in quantum physics, one cannot precisely measure two values of the same particle (for example, its position and its speed). If one of the two values are measured correctly, the other will necessarily be fuzzy. Wolfgang Pauli defines the exclusion principle, according to which two electrons can never be in the same place in the same state. Niels Bohr, of whom Heisenberg and Pauli were disciples, offers a unified theory of quantum physics.

But Albert Einstein remains very skeptical. He considers that randomness cannot be a fundamental principle of physics. If we cannot precisely measure two values of a particle, this does not mean that they are not measurable, but simply that we do not yet know how to do so. Therefore, he believes that quantum physicists use probabilities because their incomplete theory does not allow them to describe the observed phenomena correctly. Long discussions oppose Niels Bohr, during which Einstein will exclaim, "God does not play dice!"

The debate is raging.

In 1935, Einstein, Podolsky, and Rosen raised the following paradox. According to the principles of quantum physics, the state of particles is random. However, as long as they are closely linked (this is entanglement), it is possible for particles located in places very far from space to always have precisely the same state at the same time as if they were communicating instantly. These would imply that they transmit information faster than the speed of light. In the universe of Albert Einstein, as in our daily reality, it is absurd. Quantum physics must, therefore, necessarily be incomplete.

The debate remained as it was until 1964 when John Bell demonstrated in Geneva that Einstein's ideas on the incomplete aspect of quantum theory contradict this theory's very predictions. One of the other must necessarily be false. John Bell, therefore, proposed an experimental method to respond to this paradox. This experiment could not be carried out conclusively until 1982 by Alain Aspect and his Paris collaborators. It irrefutably demonstrates the correctness of the quantum theory.

Quanta

Around the end of the nineteenth century and the start of the twentieth, proof started to rise that showed that depicting light as a wave isn't adequate to represent all its watched properties. Two specific territories of the study were focal in this. The main concerns the properties of the warmth radiation transmitted by hot items. At sensibly high temperatures, this warmth radiation gets noticeable, and we portray the article as 'scorching' or, at much higher temperatures, 'emitting a white warmth'. We note that red compares to the longest frequency in the optical range, so it creates the impression that light of long frequency can be produced all the more effectively (for example at a lower temperature) than that of shorter frequency; for sure, heat radiation of longer frequencies is generally known as 'infrared'. Following the development of Maxwell's hypothesis of electromagnetic radiation and progress in the understanding of warmth (a subject to which Maxwell additionally made significant commitments), physicists attempted to understand these properties of warmth radiation. It was then known that temperature is identified with energy: the more blazing an item is, the warmer energy it contains. Likewise, Maxwell's hypothesis anticipated that an electromagnetic wave's energy ought to rely just upon its adequacy and, specifically, ought to be autonomous of its frequency. In this manner, one may expect that a hot body would transmit at all frequencies, the radiation getting more brilliant, however not evolving color, as the temperature rises. In fact, itemized counts indicated that even though the quantity of potential rushes of a given frequency increases as the frequency diminishes, shorter frequency heat radiation ought to be more splendid than that with long frequencies. Yet, again this ought to be the equivalent at all temperatures. If this were valid, all items ought to seem violet in color, their general brilliance being low at low temperatures and high at high temperatures, which isn't what we watch. This inconsistency among hypothesis and perception was known as the 'bright calamity'.

While trying to determine the bright calamity, the physicist Max Planck proposed in 1900 that the traditional laws of electromagnetism ought to be altered, so electromagnetic wave energy consistently showed up in parcels containing a fixed measure of energy. He also hypothesized that the energy contained in any of these parcels is controlled by the wave's recurrence and is more noteworthy for higher frequencies (for example, shorter frequencies). More to the point, he hypothesized that each conveyed a measure of energy equivalent to the recurrence multi-employed by a steady number that is now known as 'Planck's constant' and accepted to be a critical steady of nature; it's about 6.6×10^{-23} Js. Such a parcel of energy is known as a 'quantum' (plural 'quanta'), which is a Latin word signifying 'sum'. At moderately low temperatures, there is just enough warm energy to energize low recurrence, for example, long frequency quanta. At the same time, those of higher recurrence are created just when the temperature is higher. This is predictable with the general example of perception depicted above, yet Planck's hypothesis shows improvement over this. The equation he created on this premise produces a quantitative record of how much radiation is delivered at every frequency at a given temperature. These forecasts concur correctly with the aftereffects of estimation.

Chapter 2: Introduction to the Main Elements

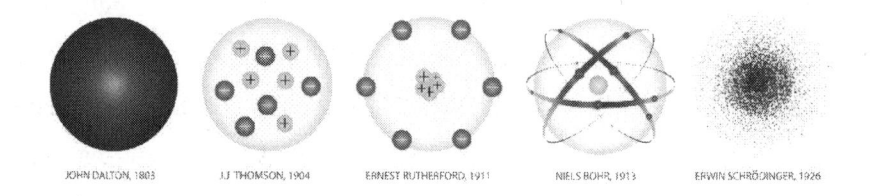

JOHN DALTON, 1803   J.J THOMSON, 1904   ERNEST RUTHERFORD, 1911   NIELS BOHR, 1913   ERWIN SCHRÖDINGER, 1926

Atomic Structure

Scientists have always tried to understand the shape and structure of an atom. A long time ago, people considered that all objects were made of indivisible things called *atoms*. But in the early 19th century, subatomic particles – particles smaller than the known atoms – were discovered! Thus, everyone wanted to develop models to explain the atomic structure.

Joseph Thomson discovered the electron. An electron is a charged particle. Thomson was working with cathode ray tubes, the ones that are used in CRT TVs. A CRT TV uses cathode ray tubes with a fluorescent screen at one of the ends. When the electron beam hits the screen, flashes of light are produced, which we see as a picture.

These glass tubes have electrodes (metallic plates) fixed on the ends. Appropriate conditions are maintained, and high voltage is applied across the electrodes (just like you connect the two ends of a battery). When Thomson used a cathode ray tube, he observed a sharp beam between the two electrodes. Thomson applied an external electric field and observed that the beam deflected towards the positive end of the external electric field, as shown below.

Thomson deduced that the beam must be made of negatively charged particles. Electrons were finally discovered. But wait! There is more... Since things around us were neutral in general, Thomson realized that the atom must also contain a positively charged particle to balance the negativity of the electron. This particle was discovered later and was named proton.

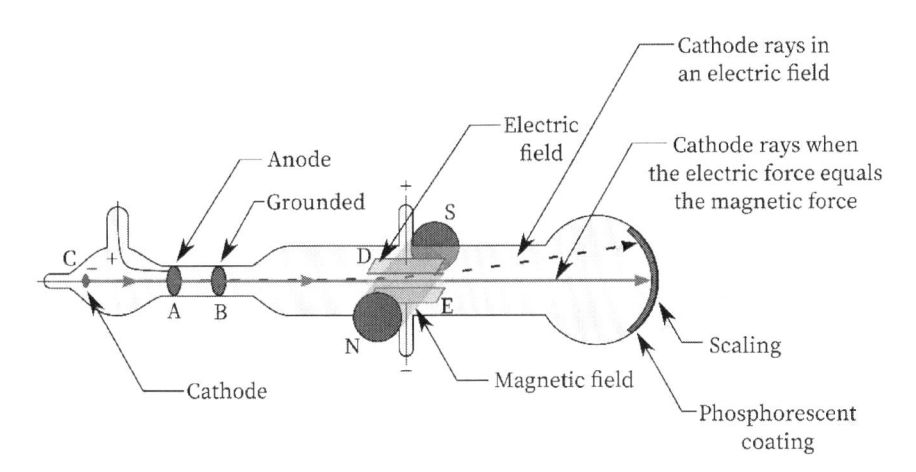

Proton was found to be much heavier than the electron because the beam made up of protons was very hard to deflect (very strong electric fields were used).

Finally, Thomson wanted to describe what an atom looked like on the inside. For that, he had to use his imagination. Since the sphere was the most *stable* object (because all water droplets were spherical, planets and stars were spherical), he deduced that atoms were also spherical.

His model is often called the *plum-pudding model.* In this model, electrons are embedded inside a positive sphere. The mass of the atom is uniformly distributed throughout the sphere. This model is shown in the image below:

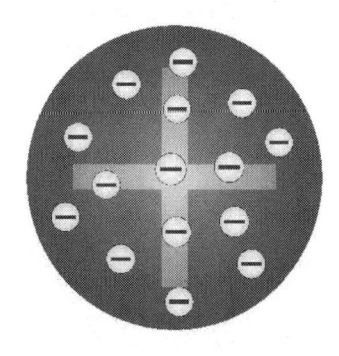

Nothing is perfect. In physics, theories are constantly being challenged and updated. Thomson's model of the atom was also tested with experiments. Rutherford, a New-Zealander, was performing advanced experiments with positively charged alpha particles and gold atoms.

With the help of his students, Rutherford designed an alpha particle experiment in which a beam of high energy alpha particles would collide with gold atoms kept at a distance. The result of the collision would be observed on a screen.

If Thomson's theory was correct, any colliding alpha particle would bounce back after the collision. It is evident that when you throw a ball towards a solid wall, it bounces back at you! According to Thomson's model, the atom is a dense sphere, but Rutherford's experiment produced one shocking result: Most alpha particles passed straight through the gold atoms, as if nothing was blocking their way!

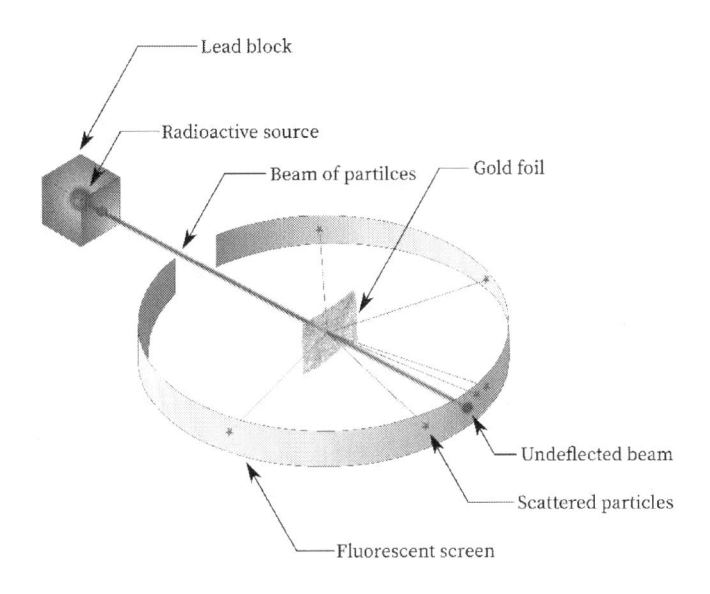

Another important observation was that 1 in 20,000 alpha particles (a tiny number) rebounded, which suggested that the alpha particle hit something really hard (dense) and bounced back! Rutherford collected all the results and created his picture of an atom. His model is often called the nuclear model of the atom.

In this model, there is a very small space at the center of an atom within which all of the mass of the atom is confined. This is called the dense nucleus. Around the nucleus, small and light-weight electrons move in their fixed orbits – just like planets revolving around the sun! A small ball of protons around which tiny electrons circle with love – interesting! The reason for this strange love is their electric field interaction. Proton and electron attract each other as they are oppositely charged particles.

Rutherford added individual masses of protons inside the nucleus to calculate the mass of an atom on paper. He did not take electrons into account because they were extremely lightweight and, thus, negligible. However, when other scientists performed experiments to determine atomic masses more accurately, they found different results – their measurement was more than what was calculated by Rutherford!

Therefore, Rutherford predicted the existence of a new subatomic particle inside the nucleus, which he called the neutron. And it had to be neutral – otherwise, it would disturb the neutrality of an atom! With this prediction, in 1908 Rutherford received the Nobel Prize in Physics.

One of his students James Chadwick, an English physicist, discovered the neutron for real! James observed neutral radiation coming out of several atoms when subjected to appropriate conditions. The neutral radiation that he observed must be composed of neutral particles – exactly what Rutherford had predicted! Chadwick also found that the mass of the neutron was almost the same as that of the proton.

Now the atom has a beautiful family:

- Electrons
- Protons
- Neutrons

However, physics is always about constant development. Even Rutherford's model was not perfect. Changes were necessary. What was the problem with his model?

Whenever a particle moves in a circular orbit, it is continuously being accelerated. This is because at every point of the circular motion, the direction of the moving particle changes. This change in direction corresponds to the changing velocity, which, in turn, corresponds to acceleration.

So, inside these atoms, electrons are under constant acceleration. But according to Maxwell's electromagnetic theory, accelerating charges emit radiation and lose energy because radiation is a form of energy and energy needs to get out of somewhere!

By the law of conservation of energy, an electron's kinetic energy (energy by its motion) should get converted into radiant energy. However, this is not observed in real life. Your body is not emitting any kind of light radiation right now! Your body is made up of atoms, remember?

If, however, the kinetic energy is getting converted into radiation, the electron must slow down eventually. And the positively charged nucleus will pull the light-weight electron towards itself – and the electron will eventually collide with the nucleus! This Rutherford picture of an atom is wrong! Atoms are not unstable like this!

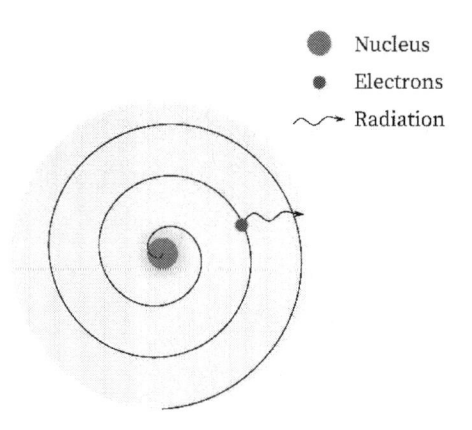

● Nucleus
● Electrons
∼→ Radiation

How do we correct this problem? Niels Bohr, the Danish physicist, came to the rescue. He used Max Planck's quantum theory and stated that electrons revolve within quantized orbits. Orbits are not randomly placed around the nucleus, as Rutherford may have assumed. "They are there for a reason," Bohr said.

These orbits can be thought of as 'fixed' energy states. Electrons, while revolving in these fixed states, cannot emit any kind of radiation. These fixed states can be calculated with mathematics. But we will not go into that right now.

"If, however, an electron absorbs energy from any outside source," Bohr said, "it will climb up the energy states." This means that there are 'fixed and unique' energy states all around the nucleus, and the further away you are from the nucleus, the more energy you possess.

But since electrons are happy only in their respective states – blame the natural laws – they want to jump back to their 'ground state'. This is all set up by nature – we develop theories thinking that nature works the same way. If the theory does not fit past observations, we make another theory and so on...

So, Bohr's theory said that orbits around the atom were not arbitrary, but stationary energy states. This model was accepted over Rutherford's model of the atom because it provided a better explanation. Rutherford's model of the atom was simple, but it was unstable. On the other hand, Bohr made only one change – that of fixed energy states. Everything else remains – the atom is mostly empty, the nucleus is small and dense, etc.

But science is all about development...

Once again, the atomic structure needed change. Even Bohr's model of the atom was not satisfactory! The new model of atom incorporates the dual nature of matter and uncertainty principle. However, we won't look any further.

Atomic Energy Level Transitions

As we have just discussed, in the early 20th century, the Rutherford model was an atomic model that was considered correct at the time. This model assumes that negatively charged electrons, like planets orbiting the sun, orbit positively charged nuclei. There are two problems with this model that cannot be solved. First, according to classical electromagnetics, this model is unstable. According to electromagnetics, an electron is constantly being accelerated during its operation. At the same time, it should lose its energy by emitting electromagnetic waves, so that it will soon fall into the nucleus. Secondly, the emission spectrum of the atom is composed of a series of discrete emission lines. For example, the hydrogen atom's emission spectrum is composed of an ultraviolet series (Ryman series), a visible series (Balmer series), and other infrared series. According to classical theory, the emission spectrum of atoms should be continuous.

In 1913, Niels Bohr proposed the Bohr model named after him. This model is the atomic structure and spectral lines and gives a theoretical principle. Bohr believed that electrons can only move in orbits with a certain energy. By absorbing photons of the same frequency, they can jump from a low-energy orbit to a high-energy one. Thus, a light quantum is released.

The Bohr model can explain hydrogen atoms, and can also explain ions with only one electron, namely He +, Li2 +, Be3 +, etc. But it cannot accurately explain the physical phenomena of other atoms.

Volatility of Electrons

As mentioned earlier, De Broglie hypothesized that electrons are also accompanied by a wave, and he predicted that electrons should produce an observable diffraction phenomenon when they pass through a small hole or crystal.

The fluctuation of electrons is also reflected in the interference phenomenon of electrons when passing through double slits. If only one electron is emitted at a time, it will randomly excite a small bright spot on the photosensitive screen after passing through the double-slit in the form of a wave. Emitting a single electron multiple times or emitting multiple electrons at a time, interference fringes between light and dark will appear on the photosensitive screen. This again proves the volatility of electrons.

The position of the electrons on the screen has a certain distribution probability, and the fringe image unique to the double-slit diffraction can be seen with time. If a light slit is closed, the image formed is the wave's distribution probability unique to the single slit.

It is never possible to have half an electron. In this electron double-slit interference experiment, it is an electron that passes through two slits simultaneously in the form of a wave. It has interfered with itself. It is worth emphasizing that the superposition of wave functions here is the superposition of probability amplitudes rather than the probability superposition as in the classic example. This "state superposition principle" is a basic assumption of quantum mechanics.

Chapter 3: Heisenberg Limited Statistics

Werner Heisenberg (1901-1976) was the prince of the theorists, so disinterested in laboratory practice that he risked flunking his thesis at the University of Munich because he did not know how batteries worked. Fortunately for him and physics as a whole, he was also promoted. There were other not easy moments in his life. During the First World War, while his father was at the front as a soldier, the scarcity of food and fuel in the city was such that schools and universities were often forced to suspend classes. And in the summer of 1918 young Werner, weakened and undernourished, was forced together with other students to help the farmers on a Bavarian farm harvest.

With the end of the war, in the first years of the twenties, we find him in the shoes of the young prodigy: pianist of high level poured in the classical languages, skillful skier and alpinist, as well as mathematician of rank lent to physics. During the lessons of the old teacher Arnold Sommerfeld, he met another promising young man, Wolfgang Pauli, who would later become his closest collaborator and his fiercest critic. In 1922 Sommerfeld took the 21-year-old Heisenberg to Göttingen, then the beacon of European science, to attend a series of lectures dedicated to the nascent quantum atomic physics, given by Niels Bohr himself. On that occasion, the young researcher, not at all intimidated, dared to counter some statements of the guru and challenge at the root of his theoretical model. However, after this first confrontation between the two was born a long and fruitful collaboration, marked by mutual admiration.

From that moment, Heisenberg devoted himself body and soul to the enigmas of quantum mechanics. In 1924 he spent some time in Copenhagen to work directly with Bohr on radiation emission and absorption problems. There he learned to appreciate the "philosophical attitude" (in Pauli's words) of the great Danish physicist. Frustrated by the difficulties to make concrete the atomic model of Bohr, with its orbits put in that way, who knows how the young man was convinced that there must be something wrong at the root. The more he thought about it, the more it seemed to him that those simple, almost circular orbits were a surplus, a purely intellectual construct. To get rid of them, he began to think that the very idea of orbit was a

Newtonian residue that had to be done without.

The young Werner imposed himself a fierce doctrine: no model had to be based on classical physics (so no miniature solar systems, even if they are so cute to draw). The way to salvation was not intuition or aesthetics but mathematical rigor. Another of his conceptual diktats was the renunciation of all entities (such as orbits, in fact) that could not be measured directly.

Measurable in the atoms were the spectral lines, the witness of the emission or absorption of photons by the atoms as a result of jumping between the electron levels. So, it was to those net, visible, and verifiable lines corresponding to the inaccessible subatomic world, that Heisenberg turned his attention. To solve this diabolically complicated problem and find relief from hay fever, in 1925 he retired to Helgoland, a remote island in the North Sea.

His starting point was the so-called "correspondence principle", enunciated by Bohr, according to which quantum laws had to be transformed without problems into the corresponding classical laws when applied to sufficiently large systems. But how big? Enough to allow to neglect the Planck constant h in the relative equations. A typical object of the atomic world has a mass equal to 10^{-27} kg. Let's consider that a grain of dust barely visible to the naked eye can weigh 10^{-7} kg: very little, but it is still greater by a factor of 10^{20}, which is one followed by twenty zeros. So, the atmospheric dust is clearly in the domain of classical physics: it is a macroscopic object and its motion is not affected by the presence of factors dependent on Planck's constant.

Heisenberg did not study in depth the most advanced frontiers of pure mathematics of his time, but he could avail himself of the help of more experienced colleagues, who immediately recognized the type of algebra contained in his definitions: they were nothing but multiplications of matrices with complex values. So-called "matrix algebra" was an exotic branch of mathematics, known for about sixty years, which was used to treat objects formed by rows and columns of numbers: matrices. Matrix algebra applied to Heisenberg's formalism (called matrix mechanics) led to quantum physics' first concrete arrangement. His calculations brought sensible results for the energies of states and atomic transitions, which is electron level jumps. When the matrix mechanics was applied not only to the hydrogen atom case but also to other simple microscopic systems, it was discovered that it worked wonderfully: the solutions obtained theoretically agreed with the experimental data. And from those strange manipulations of matrices came out also a revolutionary concept.

Suppose we indicate with x the position along an axis and p the momentum, always along the same axis, of a particle. In that case, the fact that xp is not equal to px implies that the two values cannot be measured simultaneously in a defined and precise way. In other words, if we get the exact position of a particle, we disrupt the system in such a way that it is no longer possible to know its momentum and vice versa. The cause of this is not technological, it is not our instruments that are inaccurate: it is nature that is made this way.

In the formalism of matrix mechanics we can express this idea in a concise way, which has always driven the philosophers of science crazy: "The uncertainty relative to the position of a particle, indicated with Δx, and that relative to the quantity of motion, Δp, are linked by the relation: $\Delta x \Delta p \geq \hbar/2$, where $\hbar = h/2\pi$". Said in words: "the product of the uncertainties relative to the position and the momentum of a particle is always greater or equal by a number equal to the Planck constant divided by four times pi." This implies that if we measure a position with great precision, thus making Δx as small as possible, we automatically make Δp arbitrarily large, and vice versa. You just can't have it all in life: we have to give up knowing exactly either the position or the momentum.

Chapter 4: Particle Entanglement

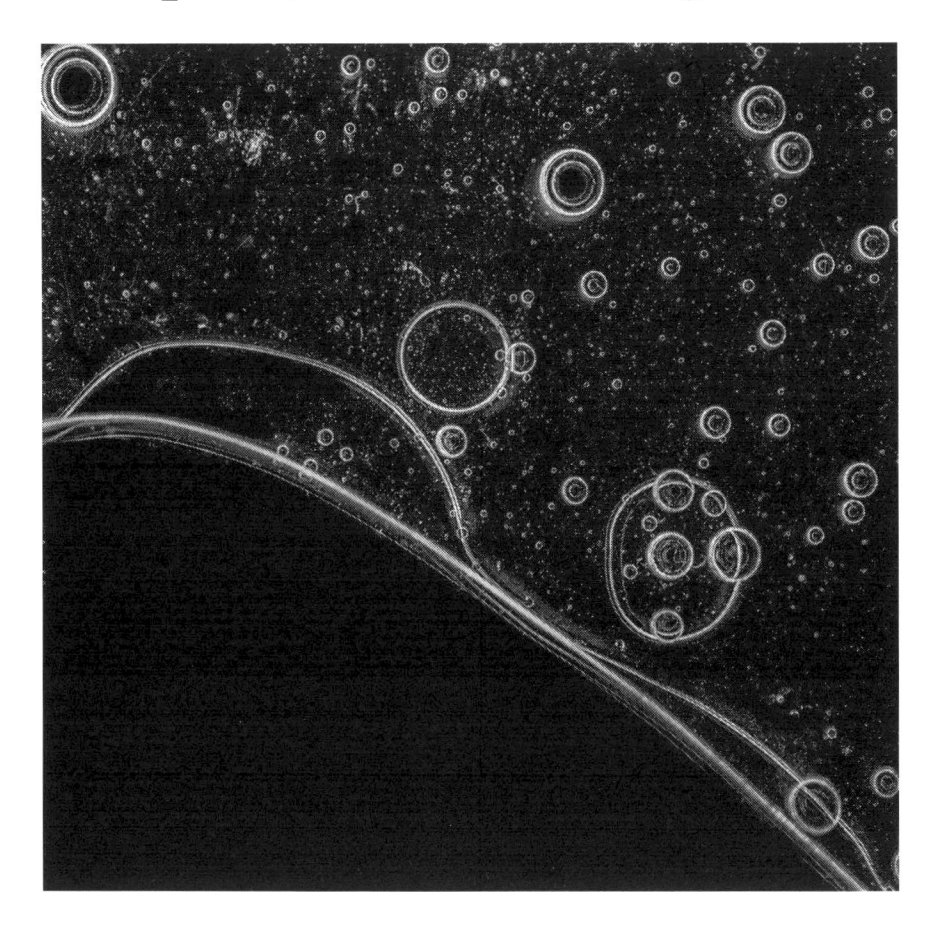

One of the basic and often seen as a strange occurrence in quantum mechanics is the concept of "entanglement." Entanglement is a phenomenon where two quantum particles interact appropriately and how their states will depend on each other, irrespective of how far apart they are.

Einstein's empirical study on momentum and position measurements led him to this concept. Indeed, entangled particles can appear remote and distant, yet they have the same physical structure, which allows an interrelation and an interconnection.

Another model to consider genuinely in our modern days is the interconnection and the polarization of the entrapped photons. It's clear that when the entanglement of electrons occurs despite the fact that particles are not connected, they are dependent upon similar movements and happenings. Concretely, this model isn't completely perceived by physicists, even though it is considered a fundamental standard of quantum physics.

The spin may have a positive or negative (upward or downward) value (or direction, known as its "sign"), and when the electrons are entangled, it means that the measurement will show that their spin signs are opposite. If the spin of one of the entangled electrons is determined, it is immediately known that the spin of the second electron has the opposite sign. In reality, this entanglement occurs, for example, when the particles are formed in a single process (in the case of experiments with photons, identical, entangled photons are produced in the process of decomposition of one photon twice their size) or when they are components of one system, such as electrons being components of one atom.

In the case of entanglement, can these characteristics be coordinated with the speed of light rather than instantaneously? There are no methods to measure time and speed with absolute precision. However, instruments' precision is constantly improving, and modern experiments have shown so far that the speed of interaction between particles during entanglement exceeds the speed of light by at least 100,000 times! Scientists assume that if the speed of interaction during entanglement exceeds the speed of light (that much), then this interaction has infinite velocity, i.e. both particles acquire the exact characteristics simultaneously regardless of distance (non-locality). For instance, if you measure two particles where they are both located in different countries and then measured at the same time, the measurement carried out in one country or location will absolutely and undeniably decide what the outcome in the other location will turn out to be.

There is no theory to describe the correlation between these particles wherein they have certain states. It is known that the two states will remain indeterminate until one of the states is weighed, it is at this time that the states of each of the particles are then determined, not minding their distances apart. A lot of experiments have been carried out in the last thirty years using atoms and light to confirm this theory. The experiment carried out now still confirms the quantum prediction.

It is worthy to note that this does not in any way serve as a means to sending signals that are faster than light, since the measurement in one location says Oxford will determine the state of the particle in another location, maybe Harvard. This shows that the outcome of each measurement is random. The particle at Harvard cannot be adjusted to match with the result obtained for the particle at Princeton. The connection between these two measurements can only be evident when the two sets of data are measured and compared, and this process has to be done at a speed which is quite slower than the speed of light.

Once a thing isn't forbidden, it is mandatory

If a quantum particle is moving from point A to point B, it will take absolutely all of the path from point A to point B simultaneously. This usually includes paths that have unlikely events like electron-positron pairs that appear from nowhere and disappear suddenly. A field of Quantum physics referred to as quantum electrodynamics (QED) implies that every possible process should be studied, including the very unlikely ones.

The QED is not just a random process of guessing without any real application.

The prediction of the interaction between the electron and the magnetic field by QED correctly describes the interaction at 14 decimal places. As strange as this theory may seem, it remains one of the best-tested hypotheses in the history of science.

Chapter 5: Angular Momentum

Basics on Angular Momentum on A Quantum Level

In quantum mechanics, the precise force administrator is one of a few related administrators similar to the traditional rakish energy. The precise force administrator assumes a focal job in the theory of nuclear material science and other quantum issues with rotational evenness. In both traditional and quantum mechanical frameworks, rakish force (together with direct force and vitality) is one of the three fundamental properties of development.

There are several operators of the angular momentum, total angular momentum (usually indicated by J), orbital angular momentum (usually indicated by L), and angular moment of rotation (in short, rotation, usually indicated by S). The term operator of the angular momentum can, confusingly, refer to the total orbital angular momentum. The whole angular momentum is always preserved.

Law of Conservation of Angular Momentum

If a skater spins on her skates, she will go faster if she forms a circle in a smaller radius. This is because a variable called angular momentum (AM, and Mom) is maintained. The equation gives the angular momentum = I * w * w. Angular velocity is a vector quantity, which implies that the localized movement direction is important for circular motion.

If a skater suddenly reduces his range of motion, his moment of inertia decreases. The speed increases to maintain the angular momentum. In other words, if it runs in a smaller circle, its speed increases. Let's take the case when traveling in a larger circle than before. The angular momentum is also maintained in this case. In this case, the moment of inertia increases. To preserve the entire AM, its speed decreases.

This law is called 'Angular Momentum Conservation Law' and applies to all objects that rotate in a circular motion.

An analogy with the same law is the law of conservation of momentum, which applies to movements in a straight line. In this case, the impulse of the colliding body system is preserved. The impulse before the collision is, therefore, MV + M1V1, where M and M1 are two different masses traveling at different speeds. In this way, the overall impulse of the system is preserved. AM and linear impulse are vectors that imply that the size depends on the direction.

Like other observable quantities, the angular momentum in QM (Quantum Mechanics) is described by an operator.

In effect, this is a vector operator, similar to the impulse operator. In short, unlike the linear momentum operator, we see the three components of the angle, and the impulse operator does not swing.

Confusingly, the term 'angular momentum' can refer to the orbit's total angle, pulse, or angular momentum.

It is possible to adopt the classical definition of the orbital angular momentum L = r × p, directly to QM by reinterpreting r and ~ p as operators associated with position e, the linear moment.

Angular Momentum Quantum Numbers

There are a lot of precise force quantum numbers related to the vitality conditions of the molecule. As far as old-style material science, rakish force is a property of a body that is in the circle or is pivoting about its hub. It relies upon the rakish speed and appropriation of mass around the hub of upheaval or pivot and is a vector amount with the course of the precise energy along with the turn hub. As opposed to traditional material science, where an electron's circle can expect a persistent arrangement of qualities, the quantum mechanical precise force is quantized.

Moreover, it can't be indicated precisely along each of the three axes all the while. For the most part, the rakish force is indicated along a pivot known as the quantization hub, and the greatness of the precise energy is constrained to the quantum esteems square root of $\sqrt{l(l + 1)}$ ($\hbar$), in which l is a whole number. The number l, called the orbital quantum number, must be not exactly the primary quantum number n, which relates to a 'shell' of electrons. Accordingly, l isolates each shell into n subshells consisting of all electrons of a similar head and orbital quantum numbers.

There is an attractive quantum number additionally connected with the rakish force of the quantum state. For a given orbital force quantum number l, there are 2l + 1 vital, attractive quantum numbers ml extending from –l to l, which confine the division of the all-out rakish energy along with the quantization hub with the goal that they are constrained to the qualities MLH, the abbreviation of Mixing Layer Height (Angular momentum). This marvel is known as space quantization and was first exhibited by two German physicists, Otto Stern and Walther Gerlach.

Basic particles, for example, the electron and the proton additionally, have a steady, natural, precise force, notwithstanding the rakish orbital energy. The electron carries on like a turning top, with its natural, precise energy of greatness s = square root of $\sqrt{(1/2)(1/2 + 1)}$ ($\hbar$), with considerable qualities along with the quantization pivot of MSH = ± (1/2) $\hbar$. There is no old-style material simple science for this purported turn, precise energy. In essence, an electron's rakish natural force doesn't require a limited (nonzero) span. However, old-style physical science requests that a molecule with nonzero precise energy must have a nonzero sweep. Electron-crash concentrates with high-vitality quickening agents show that the electron demonstrates like a point molecule down to a size of 10^{-15} centimeter, one-hundredth of the range of a proton.

The four quantum numbers n, l, ml, and ms indicate the condition of a solitary electron in an iota (a very small amount of something) totally and extraordinarily. Each arrangement of numbers assigns a particular wave work (i.e., quantum condition) of the hydrogen particle. Quantum mechanics determines how all out rakish force is developed from the precise momenta. The precise moment act as vectors to give the all-out rakish force of the molecule.

Angular Momentum on Quantum Level

The most often cited example of angular momentum is a skater that spins in a circle and draws his/her arms closer to the body and spins faster, as mentioned earlier.

But the best examples of angular momentum are the planets and stars and galaxies and satellites. For angular momentum and gravity are tied up together.

In old physics, they thought angular momentum was the following units = kg*m^2/s which is kilograms*meters^2 / seconds. That is mass*area / time. The trouble with old physics angular momentum is they forgot an important term of electric current (i).

The true equation of angular momentum needs an electric current (i) term in it.

Angular momentum = kg*m^2/(i)*s.

Do you see that extra term (i) in the denominator?

Angular momentum L = kg*meter^2/((i)*seconds) where electric current "i" is the term that every physicist of the 20th century missed.

Chapter 6: Back to Basics

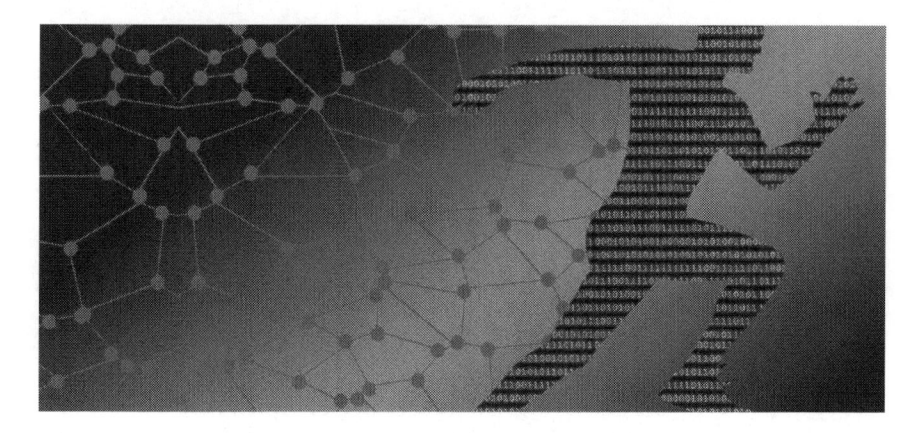

The basic mathematical framework of quantum mechanics is based on the description and statistical interpretation of quantum states, the equations of motion, the correspondence rules between observed physical quantities, the measurement postulate, and the identical particle postulate.

The world of mathematics

No doubt, you learned physics at school. What do you remember? Probably that it's always associated with mathematics. This doesn't make physics any more attractive. But unfortunately, one cannot avoid mathematics, which is indispensable for physics. Galileo Galilei (1564 - 1642), the father of modern physics, put it this way: "The book of nature is written in the language of mathematics."
Nevertheless, practically everything can be explained without it, as I do in this book.
The age-old question is whether mathematics is an invention of man

or whether it has an independent existence. That's still controversial today. But most mathematicians and nearly all physicists tend to its independent existence. This idea goes back to Plato (428 BC - 348 BC), the great Greek philosopher you most certainly know. However, he did not speak about mathematics, but rather the world of ideas and perfect forms. Which serves the universe as role models?

Is it as we know it? Undoubtedly not, because what we know is constructed by our brains. This applies both to the material world and to the world of mathematics. So, both are different in reality, but there is undoubtedly a connection. Its nature will be hidden to us for all times. Therefore, physicists and mathematicians go the pragmatic way and ignore the fundamentally unknown differences. But I think one should always keep in mind that what we call the material world and the world of mathematics is "colored" by the human brain. We only know the human versions of these worlds. Aliens, if they exist, could, therefore, have a completely different idea of both worlds.

Bohr Theory

Bohr, an outstanding contributor to quantum mechanics, states an electronic orbit quantization concept. Bohr believes that the nucleus has a certain energy level. When the atom absorbs energy, the atom transitions to a higher energy level or excited state. When the atom emits energy, the atom transitions to a lower energy level or ground state.

However, Bohr's theory also has limitations. For larger atoms, the calculation results are very inaccurate. Bohr still retains the concept of orbits in the macro world. In fact, the coordinates of electrons in space are uncertain, and electrons are more concentrated. It means that the probability of electrons appearing here is greater; otherwise, the probability is smaller. Many electrons come together and can be called an electron cloud.

Pauli's Principle

In Pauli's principle, the state of a quantum physical system cannot be completely determined, so the distinction between particles with identical characteristics (such as mass, charge, etc.) in quantum mechanics has lost its significance. In classical mechanics, each particle's position and momentum are all fully known, and their trajectories can be predicted. With one measurement, each particle can be identified. In quantum mechanics, the position and momentum of each particle are expressed by a wave function. Therefore, when the wave functions of several particles overlap each other, the practice of "tagging each particle" loses its meaning.

The indistinguishability of this identical particles has profound effects on the state's symmetry and the statistical mechanics of the multi-particle system. For example, in the state of a multi-particle system composed of identical particles, when exchanging two particles 1 and 2, we can prove that they are not symmetrical or antisymmetric. Particles in a symmetrical state are called bosons, and particles in an antisymmetric state are called fermions. In addition, the spin-exchange also forms symmetry. Particles with half of the spins (such as electrons, protons, and neutrons) are antisymmetric, so they are fermions. Particles with integer spins (such as photons) are symmetrical, so it's a boson.

The relationship between the spin, symmetry, and statistics of this esoteric particle can only be derived through the theory of relativity quantum field, and it also affects the phenomena in non-relativity quantum mechanics. One consequence of Fermions anti-symmetry is the Pauli Exclusion Principle; that is, two Fermions cannot occupy the same state. This principle has great practical significance. It means that in our material world consisting of atoms, electrons cannot occupy the same state at the same time, so after the lowest state is occupied, the next electron must occupy the next lowest state until all states are satisfied. This phenomenon determines the physical and chemical properties of the substance.

The thermal distributions of fermions and boson states are also very different. Bosons follow Bose-Einstein statistics, and fermions follow Fermi-Dirac statistics.

Bohr-Einstein Debates

The Bohr–Einstein debates is a succession of public debates on quantum mechanics that involved Albert Einstein and Niels Bohr. The importance of these debates is based on the philosophy of science and how it expanded many scientists' views. These debates occurred at a time of great change and discovery within quantum physics. While Bohr was judged the victor, it establishes the fundamental character of quantum measurement, but the scientific consensus isn't necessarily achieved.

During the 1920s, the quantum revolution occurred under both of these scientists' direction, and their debates were about understanding the changes. The uncertainty principle and various probability equations were interpretations that Einstein was unwilling to accept. His refusal to accept the complete upheaval reflected his personal desire to have a model that explained the underlying causes for the results these experiments were producing. Einstein thought there was more to be discovered. But he recognized that it could get swept under the rug of the uncertainty principle. Bohr didn't have these issues but had made peace with the apparent contradictions one could find in quantum mechanics. But Einstein moved his position over time, and his contributions to the field cannot be denied.

During what is referred to as the post-revolution period, Einstein moved through various stages that allowed him to modify his position. During his first volley, Einstein used some ingenious thought experiments to challenge the principle of indeterminacy, which he felt could be violated. His first volley was the orthodox conception of electrons and photons, during an international conference in 1927. At this argument, Einstein argues that incident particles have velocities in a practical sense that are perpendicular, and only the interaction with this deflection screen can change the original direction of propagation. Conservation of impulse implies that the sum of these impulses that interact will be conserved. Still, if the incident particle deviates at the top, the screen recoils toward the bottom, and the reverse is also true.

Bohr's answer was to explain Einstein's idea more plainly but using a mobile screen versus fixed. Bohr observed exact knowledge of the screen's vertical motion was a vital supposition in Einstein's argument. However, Bohr believed, an exact determination of the screen's velocity, when applied to the indeterminacy principle, suggests an unavoidable imprecision of its position. Before the process started, the screen would occupy an indeterminate position, to some extent. Einstein's next criticism was supposed to prove a violation occurred of the indeterminacy relation between time and energy. So, he tried to use the thought experiment designed by Bohr in 1930. Mr. Einstein expressed the idea that the existence of Bohr's experimental apparatus could be used to emphasize essential elements and key points. Using a box with electromagnetic radiation, a clock that controls the shutter's opening, and his famous E = mc2, Einstein attempted to show that, in principle, the mass of the box with electromagnetic radiation can be determined. Also, the energy within the box can be measured or determined with a precision that makes the final product less than what is implied by the indeterminacy principle.

But Bohr didn't give up. Instead, he proved that Einstein's subtle argument couldn't be conclusive, but he did so while using Einstein's idea, that of the equivalence between gravitational mass and inertial mass. Basically, the box would have to be up in the air on a spring in the middle of a gravitational field, and the weight would have to be obtained through a pointer attached to the box that links to a scale index. The unavoidable uncertainty of the box's position translated into uncertainty about the pointer's position and the weight plus energy determination. Thus, the entire thought experiment only seemed to demonstrate the uncertainty principle even more clearly. This back and forth continued into what is known as the second stage of their debate.

The second phase was regarding the orthodox interpretation characterized by the fact that it is impossible to concurrently define the values of specific discordant quantities. Einstein believed there was an ability to measure these values. So, a research line into hidden variables was done to make quantum physics complete from Einstein's point of view.

Bohr responded to Einstein's EPR paradox. It is the abbreviation of Einstein–Podolsky–Rosen paradox, as seen in the introduction. The latter occurred in a leading scientific article that contains the very first seeds of quantum physics. Bohr particularly questioned the expression that one could complete the experiment without disturbing the system in any way. With the use of possibilities and other theories of interaction, Bohr combated Einstein's paradox.

In the fourth stage of their debate, Einstein continued to refine his position, stating that quantum theory disturbed him because of its total renunciation of all minimal realism standards, even at the microscopic level. To this day, the understanding is still incomplete, and scientists continue to debate without a consensus on determinism.

Heisenberg Uncertainty Principle

One of the main differences between quantum mechanics and classical mechanics lies in the place of the measurement process in theory. In classical mechanics, a physical system's position and momentum can be determined and predicted with infinite precision. In theory, the measurement does not affect the system itself and can be performed with infinite precision. In quantum mechanics, the measurement process itself affects the system.

To describe the measurement of an observable quantity, we need to understand what an eigenstate is. The word "eigenstate" derives from the German/Dutch word "eigen" which means "inherent" or "characteristic." An eigenstate is an object's measured state, when this object possesses quantifiable characteristics such as position, momentum, etc. This object's state must be observable and must have a definite value, called an eigenvalue. Based on this, the state of a system needs to be linearly decomposed into a linear combination of a set of eigenstates of the observable quantity. It is a quantum notion which is built on a wave equation. The measurement process can be regarded as a projection on these eigenstates, and the measurement result is the eigenvalue (non-zero solution) corresponding to the eigenstate being projected. If each infinite copy of this system is measured once, we can obtain the probability distribution of all possible measured values. The probability of each value is equal to the absolute square of the coefficient of the corresponding eigenstate.

Uncertainty

The most well-known incompatible observables are the position x and the momentum p of a particle. The product of their uncertainties Δx and Δp is greater than or equal to half Planck's constant.

The "Uncertainty Principle" discovered by Heisenberg in 1927, also often referred to as "Uncertainty Relationship", refers to the mechanical quantities (such as coordinates and momentum) represented by two misaligned operators: time and energy. It is not possible to have definite measured values at the same time. The more accurate one is, the less accurate the other is. It shows that because the measurement process interferes with microscopic particles' behavior, the measurement sequence is not interchangeable, which is a basic law of microscopic phenomena.

In fact, physical quantities such as the coordinates and momentum of particles are not information that already exists and are waiting for us to measure. Measurement is not a simple reflection process but a transformation process. Their measured values depend on our measurement methods. It is the mutual exclusion of the measurement methods that leads to the uncertainty relationship.

By decomposing a state into a linear combination of observable eigenstates, the state's probability amplitude (that is a complex number used in describing the behavior of systems) can be obtained at each eigenstate. The absolute square of the probability amplitude is the probability that an eigenvalue is measured, which is also the probability that the system is in the eigenstate.

Therefore, the same measurement of an observable quantity of an identical system of the same system generally yields different results; unless the system is already in the observable quantity's eigenstate. By performing the same measurement on each system in the ensemble in the same state, a statistical distribution of measured values can be obtained. All experiments are faced with these measurement issues, and so is the statistical calculation of quantum mechanics.

Quantum Entanglement

Often the state of a system composed of multiple particles cannot be separated into the state of a single particle composed of them. In this case, the state of a single particle is called entangled. Entangled particles have amazing properties that go against common intuition. For example, the measurement of a particle can cause the entire system's wave packet to collapse immediately, so it also affects another distant particle entangled with the particle being measured. This phenomenon does not violate the special theory of relativity because you cannot define particles at the level of quantum mechanics before measuring them. In fact, they are still a whole. However, after measuring them, they will leave the state of quantum entanglement.

Quantum Decoherence and the Schrödinger's cat

As a basic theory, quantum mechanics should, in principle, apply to physical systems of any size, that is, not only limited to microscopic systems. It should provide a method for transitioning to classical macroscopic physics. The existence of quantum phenomena raises a question: how to explain the classic phenomena of macroscopic systems from the viewpoint of quantum mechanics. What can't be seen directly is how the superposition state in quantum mechanics can be applied to the macro world.

In 1954, in a letter to Max Bonn, Einstein raised the question of how to explain macroscopic objects' positioning from the perspective of quantum mechanics. He pointed out that the phenomenon of quantum mechanics is too small to explain this problem. Another example of this problem is by the Schrödinger proposed Schrödinger's cat thought experiment. However, the experiment that did not take place in the real world and was only imagined can be introduced briefly before going into its details afterward in the book. It consists of putting a cat in a small box. Then, you expose it to a radioactive substance. When the substance decays, it triggers the Geiger counter device and therefore kills the cat by the poisonous substances released by the explosion that may take place. By a double approach, the radioactive substance's decay is controlled by the binary status of 'going to decay' and 'not going to decay', which means the superposition of two paradoxical states simultaneously. In other terms, the cat is both alive and dead at the same time.

It was not until about 1970 that people began to truly understand that the thought-experiments mentioned above were not practical because they ignored the inevitable interaction with the surrounding environment. It turns out that the superimposed state is very susceptible to the surrounding environment. For example, in a double-slit experiment, the collision or emission of electrons or photons with air molecules can affect the phase relationship between the states critical to the formation of diffraction.

In quantum mechanics, this phenomenon is called quantum decoherence. It is caused by the interaction of the state of the system with the effects of the surrounding environment. This interaction can be expressed as the entanglement of each system state with the state of the environment.

The result is that superposition is only effective when considering the entire system (i.e., experimental system + environmental system). If only the system state of the experimental system is considered in isolation, only the system's classic distribution is left. Quantum decoherence is the main way to explain macroscopic quantum mechanics' classical properties in quantum mechanics today. It is also the biggest obstacle to the realization of quantum computers. In a quantum computer, multiple quantum states need to remain superimposed for as long as possible. Short decoherence time is a huge technical problem.

Chapter 7: Is it a Wave or a Particle

Waves and Particles

Numerous individuals have heard that 'wave-particle duality' is a significant component of quantum physics. We will find that at the quantum level, the results of numerous physical procedures are not unequivocally decided, and all the better we can do is to anticipate the probability or 'likelihood' of different potential occasions. We will find that something that many referred to as the 'wave work' assumes a significant role in deciding these probabilities: for instance, its quality, or power, anytime speaks to the likelihood that we would distinguish a particle at or close to that point. To gain ground, we need to know something about the wave of work properly to the physical circumstance we are thinking about. Proficient quantum physicists compute it by understanding a somewhat intricate numerical condition known as the Schrödinger condition (after Schrödinger found this condition during the 1920s); in any case, we will find that we can

Fig (a) Waves on Bondi Beach (b) Ripples on a pond

get a significant long route without doing this. Rather, we will develop an image dependent on some fundamental properties of waves, and we start with a conversation of these as they highlight in old-style physics. We all have some commonality with waves. The individuals who have lived close or visited the seacoast or have gone on a boat will know about sea waves. They can be enormous, causing severe consequences for boats, and they give a diversion to surfers when they move on to a seashore.

In any case, for our purpose, it will be more valuable to think about the more delicate waves or waves that outcome when an item, for example, a stone, is dropped into a quiet lake (Fig (b)). The figure below shows a profile of such a wave, delineating how it changes in time at better places.

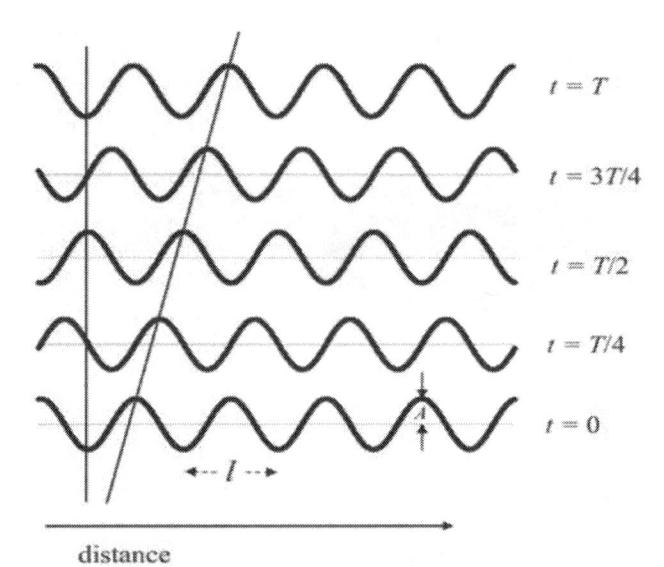

Water wave consists of a series of ripples containing troughs and peaks.

At a specific point in space, the water surface wavers here and there in a normal way. The wave's tallness is known as the 'amplitude' of the wave, and the time taken for a total swaying is known as the 'period'. Frequently it is valuable to allude to the 'recurrence' of the wave, which is the times each subsequent wave travels through a total pattern of wavering. At any moment in time, the state of the wave repeats in space, and the recurrent separation is known as the 'frequency'.

Traveling waves and standing waves

Waves, for example, those represented in the figure above, are what are called 'traveling waves' since they 'travel' in space. In the model shown, the movement is from left to right. However, it could likewise have been from right to left.

Just as traveling waves, we will need to think about 'standing waves'. A standing wave commonly happens when it is limited to a 'pit' encased by two limits. If a traveling wave is set up, it is reflected at one of the limits and moves back the other way. As a rule, the hole's dividers are with the end goal that the wave can't infiltrate them, and this outcome in the wave sufficiency is equivalent to zero at the hole boundaries. This implies just standing influxes of specific frequencies can fit into the depression – even though, for the wave to be zero at the two limits, its frequency must be the perfect length for an entire number of pinnacles or troughs to fit into the hole.

This guideline underlies the activity of numerous instruments. For instance, the note transmitted by a violin or guitar is dictated by the frequencies of the permitted standing waves on the string, which thus are constrained by the length of the string the player sets in swaying. To change the pitch of the note, the player presses the string down at an alternate point to change the length of the vibrating portion of the string. Standing waves assume a comparative job in every single instrument: woodwind and metal set up standing waves in restricted volumes of air. At the same time, the sound transmitted by drums originates from the standing waves set up in the drum skins. The kinds of sound delivered by various instruments are altogether different – even though the notes created have a distinctive 'symphonious substance'. By this, we imply that the vibration is anything but a straightforward 'unadulterated' note compared to one of the permitted frequencies, yet is developed from a mix of standing waves, the entirety of whose frequencies are products of the most reduced or 'principal' recurrence.

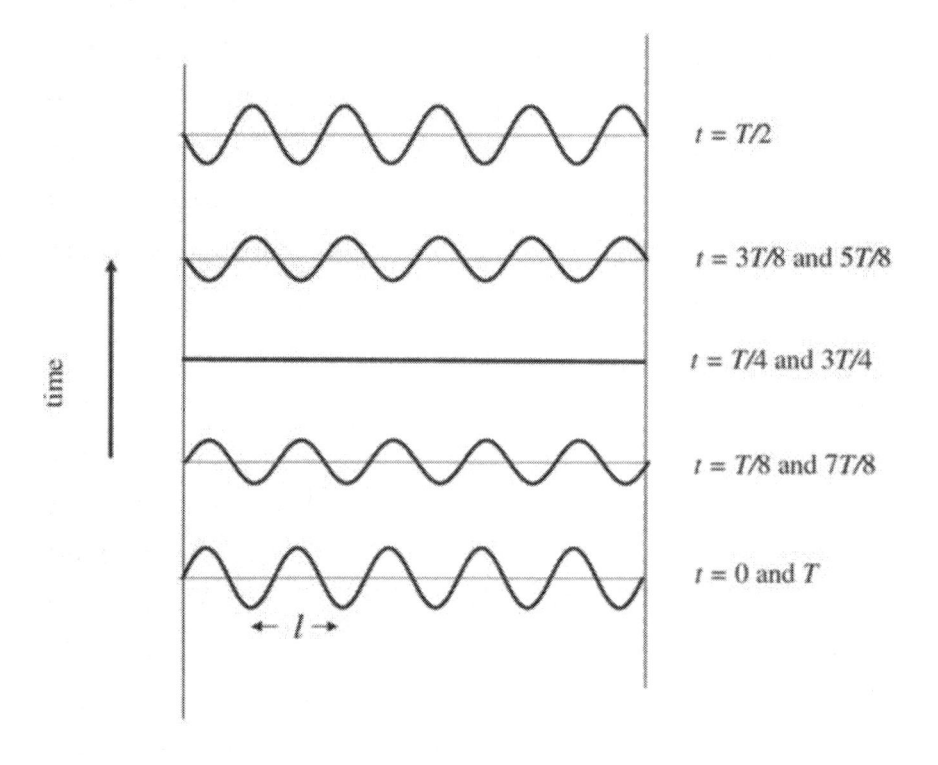

$t = T/2$

$t = 3T/8$ and $5T/8$

$t = T/4$ and $3T/4$

$t = T/8$ and $7T/8$

$t = 0$ and T

In the figure above, standing waves happen when a wave is confined to a region in space. It moves up and down in time; however, not in space.

In any case, if the standing waves were the entire story, the sound could never arrive at our ears. For the sound to be transmitted to the audience, the instrument's vibrations must create traveling waves noticeable all around, which convey the sound to the audience. For instance, in a violin, the instrument's body wavers and produces a traveling wave that transmits out to the crowd.

A significant part of the science (or specialty) of planning instruments comprises guaranteeing that the frequencies of the notes controlled by the permitted frequencies of the standing waves are replicated in the transmitted traveling waves. A full understanding of instruments' behavior and how they transmit sound to an audience is a significant point in itself, which we don't have to go into any further here. Intrigued readers could consult a book on the physics of music.

Matter Waves

The way that light, which is customarily thought of as a wave, has particle properties drove the French physicist Louis de Broglie to conjecture that different articles we generally consider as particles may have wave properties. In this way, a light emission, which is most generally envisioned as a surge of tiny slug-like particles, would, in certain conditions, act as though it were a wave. This extreme thought was first legitimately affirmed during the 1920s by Davidson and Germer: they passed an electron bar through a precious stone of graphite. They watched an obstruction design that was comparative on a fundamental level to that delivered when the light goes through a lot of cuts.

As we saw, this property is vital to the proof for light being a wave, so this test is an immediate affirmation that this model can likewise be applied to electrons. Later on, the comparative proof was found for the wave properties of heavier particles, such as neutrons. It is currently accepted that wave-particle duality is a universal property of a wide range of particles.

Indeed, even ordinary articles, for example, grains of sand, footballs, or motorcars have wave properties, even though in these cases the waves are undetectable generally – mostly because the important frequency is excessively little to be recognizable. Yet since old-style objects are made out of atoms, everyone has its related wave and every one of these waves is consistently slashing and evolving.

We saw earlier that on account of light, the vibration recurrence of the wave is straightforwardly corresponding to the energy of the quantum. On account of matter waves, the recurrence ends up being difficult to characterize and difficult to gauge legitimately. Rather there is an association between the frequency of the wave and the energy of the article. The higher the particle power, the shorter the matter-wave frequency.

In old-style waves, there is continually something that is 'waving'. Along these lines in water waves, the water surface goes all over; in sound waves, the pneumatic stress sways, and in electromagnetic waves, the electric and attractive fields change. What is the identical amount on account of matter waves? The ordinary response to this inquiry is that there is no physical amount that compares to this. We can ascertain the wave using quantum physics' thoughts and conditions, and we can use our outcomes to foresee the estimations of amounts that can be estimated tentatively. However, we can't legitimately watch the wave itself, so we need not characterize it genuinely and should not endeavor to do so. To stress this, we use the term 'wave work' instead of a wave, which underlines the point that it is a numerical capacity as opposed to a physical article. Another significant specialized contrast between wave capacities and the traditional waves we talked about before is that while the old-style wave wavers at the wave's recurrence, in the matter-wave case, the wave work stays consistent in time. Notwithstanding, because not physical in itself, the wave work assumes a basic job in quantum physics to understand genuine physical circumstances. Right off the bat, if the electron is limited to a given district, the wave work forms standing waves like those talked about before; subsequently, the frequency and along these lines the particle's energy takes on one of a lot of discrete quantized qualities. Furthermore, if we do trials to identify the nearness of the electron close to a specific point, we are bound to discover it in areas where the wave work is enormous than in ones where it is little. This thought was set on an increasingly quantitative premise by Max Born, whose standard expresses that the likelihood of finding the particle at a specific point is relative to the block of the extent of the wave work by then.

Atoms contain electrons bound to a little district of the room by the electric power pulling in them to the core. From what we said before, we could expect the related wave capacities to frame a standing-wave example, and we will see in a matter of seconds how this prompts an understanding of significant appropriate ties of particles. We start this conversation by considering a less complex system where we envision an electron to be limited to a little box.

An electron in a container

In this model, we think about the instance of a particle, which we will accept to be an electron, caught inside a crate. By this, we imply that if an electron is in the container, its potential energy has a steady worth, which we can take to be zero. The electron is bound to the case since it is encircled by a locale of exceptionally high potential energy, which the electron can't enter without breaking the rule of energy protection. An old-style similarity would be a ball inside a block box lying on the floor: given the sides of the container are sufficiently high, the ball can't escape from the case, because to do so it would need to defeat gravity. Before long, we will be thinking about the matter waves proper to this circumstance, and we may contrast these with the instance of a lake or pool, where a strong fringe encircles the water. The strong shore is unequipped for vibrating, so any waves created must be restricted to the water.

As a supplementary improvement, we treat the issue as 'one-dimensional', by which we imply that the electron is bound to move along a specific heading in space with the goal that movement in different ways can be overlooked. We would then be able to make a similarity with waves on a string, which are one-dimensional because they can just move along the string. We now consider the type of electron wave work. Since the electron can't escape from the case, the likelihood of discovering it outside is zero. If we think about the crate's very edge, the likelihood of finding the particle by then can have just one worth, so the way that it is zero fresh implies that it should likewise be zero simply inside. This condition is exceptionally similar to that applied to a violin or guitar string. We saw before, this infers the wave must be a standing wave with frequency to such an extent that it fits into space accessible. What is shown in this implies just these specific estimations of the frequency are permitted, and, as the frequency dictates the electron power through the de Broglie connection, the energy is likewise limited to a specific arrangement of qualities. Recalling that the potential energy is zero and that the electron's motor energy relies just upon its (known) mass and its power, we see that the total energy is also restricted to one of a lot of specific qualities – for example, the energy is 'quantized' into a lot of 'energy levels'.

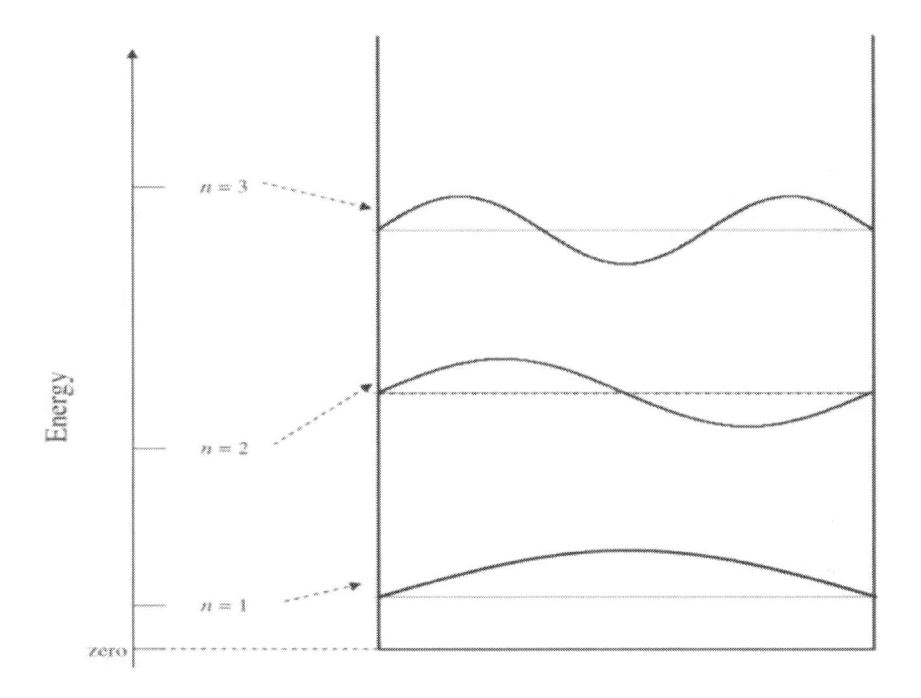

In this figure, the energy levels and waves function for an electron's energy states in an electron box.

Readers may well have gone over the reference to the 'Heisenberg vulnerability standard'. This is named after Werner Heisenberg who, as we have seen, was a pioneer of quantum physics and formulated his way to deal with the subject quickly before Schrödinger built up his condition. When all is said in done terms, the vulnerability guideline expresses that it is difficult to know the specific estimations of two physical amounts, for example, the position and energy of a particle simultaneously. We can perceive how this function works by alluding to our case of the particle in the container. If we initially think about its position, all we know is that the particle is someplace in the case. We characterize the vulnerability in position as the good ways from the inside to the case edge, which is a large portion of the crate size. Going to energy, if we consider a particle in the ground state, the wave work has the type of part of a wave whose frequency is double the crate size: as the particle could be moving in either heading (left or right), the vulnerability in power (characterized comparably to that in position) is its greatest size, which relies upon the frequency. If the case were bigger, the uncertainty in position would be bigger, and however, the energy would be littler. If we duplicate these amounts together, we find that the crate size counterbalances and the item rises to Planck's steady. The Heisenberg vulnerability guideline expresses that the result of the vulnerabilities in position and energy can never be littler than a number roughly equivalent to one-tenth of Planck's steady. We see this is undoubtedly the situation for our model. This is a general property of any wave work related to a quantum state; we should take note that the vulnerability standard is subsequently an outcome of wave-particle duality and, accordingly, quantum physics, as opposed to something extra to it.

If we had various indistinguishable boxes containing electrons, their ground states would likewise be indistinguishable. One of the properties of particles that we were unable to clarify traditionally was that all atoms of a given sort have similar properties, and specifically that they all have the equivalent most reduced energy state. Through wave-particle duality, quantum physics has clarified why such a state exists on account of an electron in a container, and we will see in a matter of seconds how similar standards apply to an electron in an atom.

Let's remember that momentum equals mass times velocity and that the potential energy is zero, in this case, the energy of the particle in a box is:

$E = mv^2 = p^2/2m = (h^2/8\, mL^2)\, n^2$ (where h is the Planck's constant and L the length of the string).

If L is similar to the size of an atom (say 3×10^{-10} m), then, using the known value of the mass of an electron (m = 10^{-30} kg), E = $5 \times 10^{-19}\, n^2$ J.

The change in energy when an electron moves from its n = 2 to its n = 1 state is $3\, h^2/8\, mL^2 = 1.1 \times 10^{-18}$ J.

If this energy is given to a photon, the frequency, f, of the associated electromagnetic wave will be this divided by h, and the corresponding wavelength is:

$l = c/f = 8\, mL^2\, c/3h = 1.1 \times 10^{-7}$ m

This is quite similar to the wavelength of the radiation emitted when a hydrogen atom makes a transition from its first excited state to its ground state, which is 1.4×10^{-7} m.

Now consider what occurs to the electron in a case when it changes starting with one permitted energy level then onto the next state from the main energized state to the ground state. To ration energy, the energy lost must head off to someplace, and if we accept it is discharged as a quantum of electromagnetic radiation, the frequency of this radiation can be determined from the distinction between the energy levels using the Planck formula. Again, we perceive how quantum physics represents nuclear properties that we couldn't clarify traditionally.

We should be assured that the numbers come out about the correct size, and we can at least likely accept that a few properties of atoms result from the wave idea of their electrons. In any case, we should recollect that there are quite significant contrasts between a genuine three-dimensional particle and our one-dimensional box. We realize that particles consist of contrarily charged electrons pulled into a decidedly charged core with the goal that the possible energy of fascination reduces the further the electron is from the core.

The outcome is to bind the electron to the region of the core, and we could expect the wave capacities to be standing waves. In any case, not exclusively in the nuclear 'box' three-dimensional. Yet, its shape is not quite the same as that discussed above, so we may not be completely persuaded of our methodology's rightness before we have applied it to the genuine nuclear potential. We will come back to this in more detail in the blink of an eye.

Shifting potential energy

So far, we have considered the matter waves related to particles proliferating in free space or caught in a one-dimensional box. In both these cases, the particle moves in an area where the potential energy is steady; in this way, if we remember that the complete energy is monitored, the active energy and thus the particle's power and speed must be the equivalent of any place it goes. Interestingly, a ball moving up a slope, for instance, increases likely energy, loses dynamic energy, and eases back down as it climbs. Now we realize that the de Broglie connection connects the particle's speed to the frequency of the wave, so if the speed remains steady, this amount will likewise be the equivalent all over the place, which is the thing that we have verifiably accepted. However, if the speed is inconsistent, the frequency should likewise differ, and the wave won't have the generally straightforward structure we have considered up until now. Therefore, when a particle travels through an area where the potential energy fluctuates, its speed and consequently the wave capacity's frequency will likewise change.

Generally, the investigation of a circumstance where the potential energy differs requires an investigation of the scientific condition that controls the type of wave in the general case. As referenced before, this condition is known as the 'Schrödinger condition'. In the models discussed above, where the potential is uniform, the Schrödinger condition arrangements have the type of traveling or standing waves, and our genuinely basic methodology is defended. A full understanding of more broad circumstances is scientifically very testing and not fitting to this book. Nevertheless, we can increase a great deal of understanding based on our prior conversation if we are set up to accept a portion of the subtleties based on previous experience. In a matter of seconds, we will apply this to an investigation of the structure of atoms. In the following section, we will see that straightforward traveling and standing waves can speak to electrons' movement in metals. In the first place, however, we will attempt to extend our understanding of the wave idea of a particle moving in a differing potential by thinking about two further models.

Quantum tunneling

We initially consider the instance of a particle moving toward a 'likely advance'. We are especially inspired by the situation where the energy of the drawing closer particle is littler than the progression stature. So from an old-style perspective, we would anticipate that the particle should ricochet back when it arrives at the progression and then move in reverse at a similar speed. Much something very similar happens when we apply quantum physics. However, there are significant contrasts, as we will see. First, we think about the type of matter-wave. Based on our previous conversation, we expect particles moving toward the progression to be represented by traveling waves moving from left to right. At the same time, after they skip back, the wave will head out from the option to left. Generally, we don't know what the particle is doing at a specific time, so the wave capacity to one side of the progression will be a mix of these, which is affirmed when the Schrödinger condition is tackled scientifically. What is of genuine intrigue is the type of wave to one side of the progression.

Traditionally, there is no likelihood of finding the particle there, so we may expect the wave capacity to be zero in this district. In any case, when we explain the Schrödinger condition, we find that the power of the wave work anytime speaks to the likelihood of finding a particle by then. We see that quantum physics predicts a limited possibility of discovering it in an area where it would never be if old-style physical science were the entire story. It ends up being difficult to test the above forecast legitimately, since setting any sort of finder inside the hindrance would successfully change the type of the potential. When the Schrödinger condition is fathomed in this circumstance, we find that the type of the wave capacity to one side of the hindrance and inside it is fundamentally the same as that just talked about on account of the progression. However, there is currently a traveling wave of nearly little, yet limited, sufficiency to one side of the obstruction. Deciphering this truly, we presume that there is a little likelihood that a particle moving toward the boundary from the left won't bob back yet will rise to the opposite side. This marvel is known as 'quantum mechanical tunneling' because the particle seems to burrow through an obstruction that is invulnerable traditionally.

In the next page figure, the straight lines in (a) represent a potential step. The wave function for a particle approaching the step is also shown; it penetrates the step, giving a probability of finding the particle in a region that is forbidden classically. The corresponding case for a narrow barrier is shown in (b) the wave function penetrates the barrier so that there is a probability of the particle emerging on the right-hand side, where it could never be classically. This is known as 'quantum mechanical tunneling'.

(a)

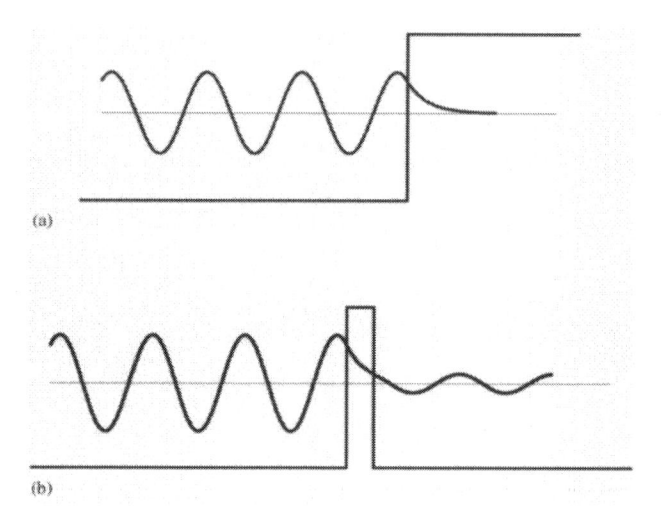

(b)

There is a wide scope of physical wonders that demonstrate quantum tunneling practically speaking. For instance, in numerous radioactive rots, where 'alpha particles' are discharged from the cores of certain atoms, the likelihood of this incident for a specific particle is probably low – so low in the certainty that a particular core will hold up a huge number of years on normal before rotting. This is currently understood on the premise that the alpha particle is caught inside the core by what might be compared to a likely obstruction, comparative on a basic level to that talked about above. A low abundance wave exists outside the obstruction, which implies that there is a little (however non-zero) likelihood of the particle tunneling out.

In the next page, figure (a) shows a scanning tunneling microscope that moves a sharp point across a surface and detects the tunneling current into the surface. This varies strongly with the distance of the point from the surface, so any unevenness can be detected. Figure (b)

(a)

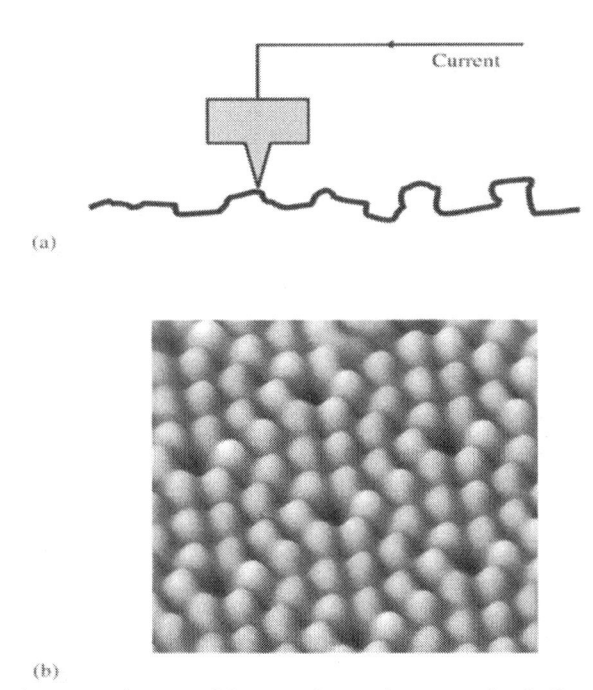

(b)

shows an image of part of the surface of a crystal of silicon; the bright peaks correspond to individual atoms.

Recently, quantum tunneling has been significantly misused in the filtering tunneling magnifying lens. In this gadget, a sharp metal point is held simply over a metal surface. Subsequently, electrons burrow through the hindrance isolating the metal point from the surface and a current stream. Suppose the fact of the matter is now examined over a lopsided metal surface. In that case, the tunneling current variations give information about this lopsidedness and a guide of the surface outcomes. Researchers' capacity to watch and control singular particles using filtering tunneling microscopy and other comparable methods has opened up an entirely different science and innovation field known as 'nanoscience'.

A quantum oscillator

In the traditional case, the particle would waver consistently from one side of the potential well to the next with a recurrence dictated by its mass and the well's state. The size, or 'sufficiency', of the wavering, is controlled by the particle's energy: at the foot of the well, this energy is motor, while the particle stops at the constraints of its movement, where all the energy is potential. The wave capacities are obtained by fathoming the Schrödinger condition, and it is discovered that similarly, as on account of a particle in a crate, standing-wave arrangement is conceivable just for specific estimations of the energy. First, the similitudes: in the two cases, the wave work corresponding to the most reduced energy state is spoken to by a single protuberance that arrives at a greatest in the inside; the following most noteworthy state has two mounds, one positive and the other negative with the wave work crossing the hub, etc. Now the distinctions.

In the figure in the next page, the energy levels and wave functions correspond to the four lowest energy states of a particle moving in a parabolic potential. The wave functions have been drawn so that their zeros are at the corresponding energy levels.

In the first place, the width involved by the wave is the equivalent for all states on account of the crate. Yet, in the oscillator case, as the absolute energy increases, so does the width of the area where the total energy is sure. Generally, we can say that the powerful width of the container is distinctive for the diverse energy levels. Furthermore, the wave doesn't go to zero quickly. The constraints of the old-style movement are reached. However, it infiltrates the 'traditionally illegal' area somewhat in a way like an instance of an atom moving toward a stage.

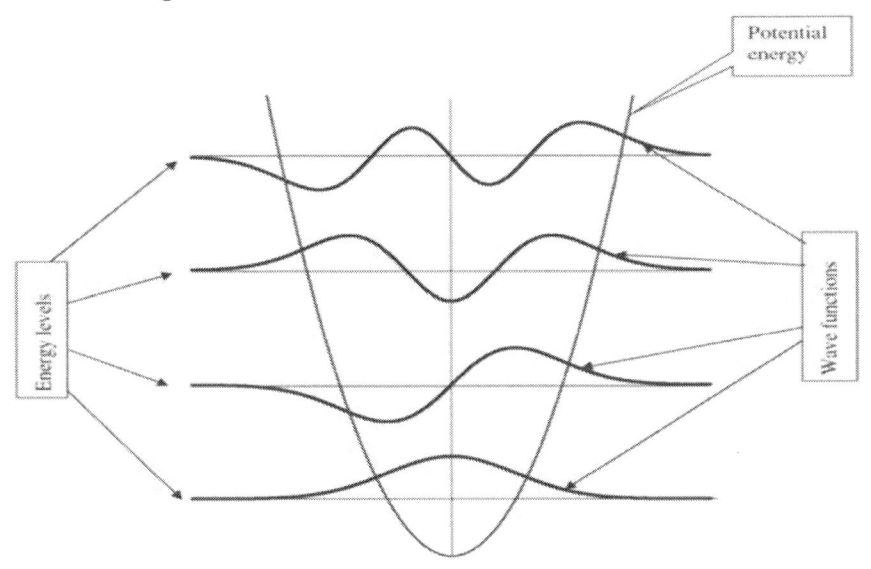

By considering this model, the reader will ideally have acknowledged what number of the highlights of such an issue can be derived from the understanding of matter waves in a consistent potential, even though the subtleties require an increasingly numerical methodology. We will currently attempt to apply these standards to see a portion of genuine atoms' quantum physics.

The hydrogen atom

The most straightforward particle is that of the component hydrogen, which comprises of a single contrarily charged electron bound to an emphatically charged core by the electrostatic (or 'Coulomb') power, which is solid when the electron is near the core and consistently decreases in quality when the electron is further away. Therefore, the potential energy is enormous and negative, close to the core, and draws nearer to zero as we move away from it. The models examined so far have all been one-dimensional, implying that we have verifiably accepted that the atom is compelled to move along a specific course (from left to right or the other way around in our graphs). However, particles are three-dimensional articles, and we should consider before we can understand them completely. A significant improving element of the hydrogen atom is that the Coulomb potential is 'roundly symmetric' – for example, it relies just upon the separation between the electron and the core – whatever the heading of this division. A result is that a significant number of the wave capacities related to the permitted energy levels have a similar balance; we will examine these first and come back to the others later.

The Coulomb potential limits the electron to the core region as the block box, and the oscillator potentially restricts the atom in the models discussed above. We perceived how the successful box width in the oscillator case was bigger for higher energy conditions, which implied that the energies of the higher states didn't increase as quickly as on account of the block box. Applying similar thinking as in the oscillator case, we could expect the energy levels to increase much more gradually as we go up the stepping stool. This is for sure what occurs, and the energy levels come out to be − R, − R/4, − R/9, − R/16 ... where R is a consistent known as the 'Rydberg' steady, after the Swedish researcher who dealt with nuclear spectra towards the end of the nineteenth century. Notice that these numbers are negative since we measure the energy from a zero level that compares to the electron and core being exceptionally far separated.

When an atom moves starting with one energy level, then onto the next, the energy is assimilated or discharged as a photon of radiation, whose recurrence is identified with the energy change by the Planck connection. The example of frequencies determined along these lines from the above example of energy levels is equivalent to that watched when electrical releases are gone through hydrogen gas. Thus, we now have a total quantitative understanding between the expectations of quantum physics and the exploratory estimations of the energy levels of the hydrogen atom.

We have used the rule of wave−atom duality to get the quantized energy levels, however, how are we to decipher the wave work that is related to each level?

The response to this inquiry lies in the born standard expressed before. The block of the wave work anytime speaks to the likelihood of finding the electron close to that point. A model of the atom steady with this is, in this unique circumstance, that the electron could be thought of not as a point atom, yet as a constant distribution spread over the volume of the atom. We can visualize the atom as a decidedly charged core encompassed by a haze of negative charge whose focus anytime is proportional to the wave work block by then. This model functions admirably as a rule; however, it should not be taken too truly. If we search for the electron in the atom, we will consistently discover it as a point atom. Then again, it is similarly off base to think about the electron just like a point atom when we are not watching its position. In quantum physics, we use models, yet don't decipher them too. We come back to this in our conversation about the theoretical standards of the subject later.

So far, we have examined just states where the wave has around balance: for example, it has a similar incentive in similar good ways from the core, whatever the course. In any case, different states don't have this straightforward property. However, they shift in the heading. The physical essentialness of these non-circular states is that they relate to the electron moving around the atom with some 'precise speed' and related 'rakish energy'.

Conversely, the circular waves compare to the electron being spread around the volume of the core, however having no orbital movement. Given that there are, on the whole, these non-round states, why isn't the range of energy levels substantially more mind-boggling than we have talked about? By fortunate happenstance, for reasons unknown, the energy of each of the non-circular states is equivalent to that of one of the round states, so the straightforward picture we examined before holds. If it had not been for this upbeat mishap, the trial range of hydrogen would not have fitted the nearly basic equation talked about above, and the way to an effective quantum hypothesis would have been significantly harder.

Chapter 8: How Light Behaves

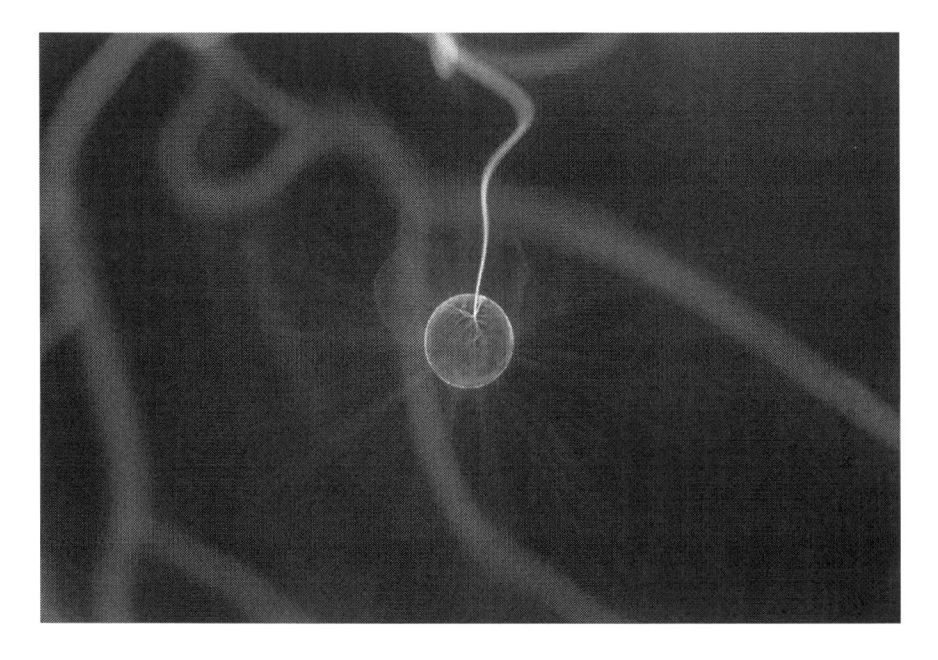

Light behaves like a wave in one moment and a particle in the future. This particle, called a photon, was verified by Einstein through his experiments on the photoelectric effect. In this experiment, he concluded that energy packets are released and that this packet is the photon. The tests also confirmed the existence of photons.

For example, in an experiment, the light was passed through thin slits like wafers, and on the other side, there was a film. Whenever a photon hit the film, it left traces in the film, and this continued indefinitely until a very interesting vibration emerged. Photons repeatedly hit the same areas avoiding other areas of the film.

The outcome was a progression of light and dull lines on the film. This was a wave theme, even though it was made by the photons hitting the film. The molecule had shaped a wave design. Different investigations have checked the wave idea of light.

Light can be gone through a crystal and separated into various shading frequencies. For these reasons, we accepted the fact that light behaves both like a wave and like a particle. Experiments will show it is a particle, and other analyses will show it is a wave.

The light is very different from the other waves we know. Light is nothing but energy, which consists of an electric wave and a magnetic wave. These two waves move perpendicular to each other and perpendicular to the direction of movement. For this reason, light is called an electromagnetic wave.

The most ordinary waves we know, such as a wave on a lake or in the ocean are the energy that moves through a mass. If you dropped a stone into a pond, you would see a ripple moving outward and away from where the rock entered the water. Now, the motor vitality of the rock was consumed by the lake when the stone hit its surface. This energy then moves across the surface of the pond. It is important to know that water does not run. What you see is the wave moving through the water while the water remains in the same place. Only the wave made the journey. This is not only the case with liquid bodies.

All forms of matter can absorb energy. Take a spoon and tap lightly on a glass. You will not see a wave passing through the glass, but you will feel it, and part of the energy of the impact will pass through the glass. Now hit the glass a little harder, and this time it will break. Rigid glass is unable to process the amplitude of the energy you have applied, and it breaks.

A superior model is thinking about quakes. A seismic tremor is just a prompt and abrupt arrival of putting away vitality. The pressure builds up over time on a fault line until it is released. When released, energy from this structure is also released. The resulting earthquake is a wave that crosses the earth's crust. All these events I have described are energy that flows through a medium in the form of waves. This is not the case with light.

Light is an electromagnetic wave. How does this wave behave, and how is it structured? As I said, light acts both as a particle and as a wave. This says a lot about the light because we have the parts, and it's just to solve the puzzle. When light acts as a particle, we know that a particle must be present. If light acts like a wave, we know there must be a wave.

It would be far-fetched to believe that light behaves in a certain way, depending on how we look at it. The light does not say, "Now that they're experimenting with me like that, I should act like a particle." The light does not decide anything. It is what it is, and we must be able to read the clues and formulate the correct conclusions from those clues.

The outline below shows the necessary rules for the existence of light as we know it:

1) The presence of electric and magnetic waves is required for the survival of the photon.
2) The speed of light must remain above a certain threshold; otherwise, the wave idea of the light will be pulverized.
3) The photon can't exist without the electromagnetic wave segment.

4) The light will bend in the presence of a strong magnetic field.

Consequently, as long as the light moves at high speed, it shows wavy properties. When the light drops below the speed required to keep the wave intact, the light wave decays into a stream of free photons. This is because magnetic and electric waves are responsible for the existence of the photon.

The magnetic component of the light wave detects the presence of magnetic fields and reacts in the presence of this field. This refers to the theory that gravity is a manifestation of magnetism on a macro scale. We observe magnetism on a micro-scale daily. The macroscale includes large mass bodies such as moons, planets, stars, solar systems, galaxies, *etc*. It is an amount of space that contains a large number of particles, which together create a strong gravitational field.

Light waves

Light waves are not the same as water waves and sound waves in that nothing is compared to the vibrating medium (for example, the water, string, or air) in the models talked about before. To be sure, light waves are fit for going through void space, as is evident from the way that we can see the light discharged by the sun and stars. This property of light waves introduced a significant issue to researchers in the eighteenth and nineteenth hundred century. Some inferred that space isn't unfilled yet loaded up with a hidden substance known as 'aether,' which was thought to help the swaying of light waves. In any case, this speculation ran into inconvenience when it was understood that the properties required to help the exceptionally high frequencies average of light couldn't be accommodated as the aether offers no protection from the development of objects (for example, the Earth in its circle) through it.

Around then, the physics of power and attraction was being created, and Maxwell had the option to show that it was contained in a lot of conditions (now known as 'Maxwell's conditions'). He additionally demonstrated that one kind of answer for these conditions compares to the presence of waves that comprise wavering electric and attractive fields that can go through void space without requiring a medium. The speed these 'electromagnetic' waves travel at is dictated by the critical constants of power and attraction. When this speed was determined, it was seen as indistinguishable from the deliberate speed of light. This drove legitimately to the possibility that light is an electromagnetic wave, and it is now realized that this model likewise applies to a scope of other marvels, including radio waves, infrared radiation (warmth), and X-beams.

Interference

Direct proof that a wonder, for example, light, is a wave, is acquired from contemplating 'interference'. Interference is usually experienced when two influxes of a similar frequency are included. Interference is critical proof for light's wave properties, and no other traditional model can represent this impact. Assume, for instance, that we rather had two surges of old-style particles: the total number of particles would consistently approach the whole of the numbers in the two bars, and they could always be unable to offset each other in the manner that waves can.

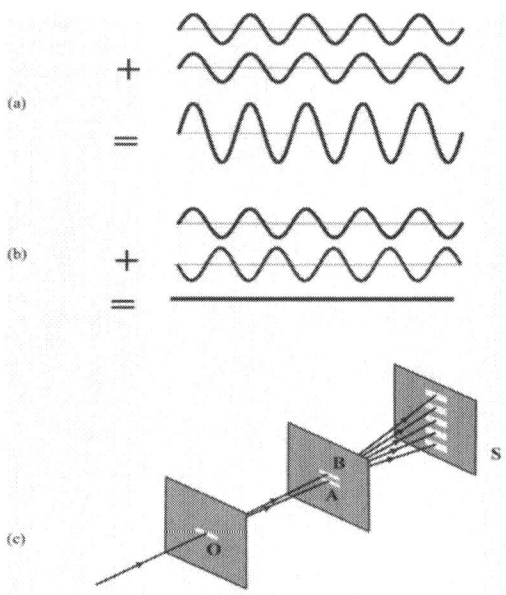

When two waves in step are combined, they reinforce each other like (a), but if out of step with each other, they cancel each other like (b). (c) explains Young's experiment.

The primary individual to watch and clarify obstruction was Thomas Young, who, around 1800, played out an investigation like that outlined in Figure (c). The light goes through a restricted cut named O, after which it experiences a screen containing two cuts, A and B, lastly arrives at a third screen, S, where it is watched. The light arriving at the last screen can have gone by one of two courses – either by A or by B. In any case, the separations went by the light waves following these two ways are not equivalent, so they don't generally show up at the screen in sync with one another. It follows from the conversation in the past passage that at certain points on S, the weaves will strengthen one another, while at others, they will drop; therefore, an example consisting of a progression of light and dim groups is seen on the screen.

Light Quanta

In 1905, Albert Einstein (around then obscure to mainstream researchers) distributed three papers that were to revolutionarily affect the fate of physics. One of these identified with the marvel of 'Brownian movement', in which dust grains in a fluid are believed to move indiscriminately when seen under a magnifying instrument: Einstein demonstrated this was because of them being shelled by the particles in the fluid and this knowledge is commonly perceived to establish the last proof of the presence of atoms. Another paper (the one for which he is generally commended) set out the hypothesis of relativity, including the popular connection between mass and energy. However, we are worried about the third paper – for which he was granted the Nobel Prize for physics – which offered a clarification of the photoelectric impact dependent on Planck's quantum theory. Einstein understood that if the energy in a light wave is conveyed in fixed quanta, then when light strikes a metal, one of these will move its energy to an electron. Subsequently, the energy conveyed by an electron will be equivalent to that conveyed by a light quantum, less a fixed sum required to expel the electron from the metal (known as the 'work'), and the shorter the frequency of the light, the higher will be the energy of the discharged electron. When estimations of the properties of the photoelectric impact were broken down on this premise, it was discovered that they were in concurrence with

Einstein's theory and the estimation of Planck. What was consistently found from these estimations was equivalent to that obtained by Planck from his investigation of warmth radiation.

A significant extra perception was that, regardless of whether the power of the light is feeble, a few electrons are transmitted promptly, the light is turned on, suggesting that the entire quantum is immediately moved to an electron. This is exactly what might occur if light were made out of a flood of particles as opposed to a wave, so the quanta can be thought of as light particles, which are called 'photons'. In this manner, we have proof from the impedance estimations that light is a wave, while the photoelectric impact shows that it has the properties of a flood of particles. This is what is known as 'wave-particle duality', as we have discussed earlier. A few readers may expect, or if nothing else, trust that a book like this will disclose to them how light can be both a wave and a particle. In any case, such a clarification presumably doesn't exist. The wonders that display these quantum properties are not part of our consistent experience (even though it is a significant point of this book to show that their outcomes are) and can't be completely depicted using old-style classes, for example, waves or particles, which our brains have advanced to use. Indeed, light and other quantum objects are once in a while totally wave-like nor completely particle-like, and the most proper model to use for the most part relies upon the experimental setting. When we play out an impedance, try different things with exceptional light emission, we, for the most part, don't watch the behavior of the individual photons, and to a generally excellent approximation, we can speak to the light as a wave. Then again, when we distinguish a photon in the photoelectric impact, we can helpfully consider it a particle. These depictions are approximations in the two cases, and the light joins the two viewpoints to a more noteworthy or lesser degree. Endeavors to understand quantum questions all the more deeply have raised theoretical difficulties and prompted incredible philosophical discussions in the last hundred or so years. Such contentions are not key to this book, which plans to investigate quantum physics outcomes for our ordinary experience. The second arrangement of marvels that prompted the quantum to hypothesize is known as the 'photoelectric impact'. When light strikes a perfect metal surface in a vacuum, electrons are discharged. These all convey a negative electric charge, so the flood of electrons comprises an electric flow. Applying a positive voltage to the metal plate can stop this current, and the littlest voltage that can do so gives a proportion of the energy conveyed by every electron. When such analyses are done, it is discovered that this electron energy is consistently the equivalent for the light of a given frequency. If the light is made more brilliant, more electrons are radiated. However, the energy conveyed by every individual electron is unaltered.

Chapter 9: The Theory of Relativity

In 1907, only two years after developing the theory of special relativity, Einstein had the idea that he would describe as "the happiest of his entire life." In this inner vision, what would be revealed as the essential physical basis of general relativity appeared to him, even if it would take him almost ten years to elaborate the theory mathematically. Einstein realized that "if a man falls freely, he would not feel his weight."

Even the expression "free fall" is telling though, that one is always attached to a gravitational field, attracted to the Earth from Newtonian theory's perspective. One finds freedom when one is falling. It is this freedom that those who pursue free-falling as a hobby seek to find and to feel, even if it is only partly due to air resistance. It is, of course, astronauts in "weightlessness" who truly experience over a long period this feeling of no longer having any weight, of no longer being subject to the force of the Earth's attraction. Nevertheless, Einstein's great idea was the understanding that, if we jump up, during the brief moment of our jump, we experience this "weightlessness."

In more words, there is no difference in principle between a vessel in orbit around the Earth and a ball which we throw here on Earth: both are in free fall; both are, for the duration of their motion, satellites of the Earth.

The Equivalence Principle

Understanding this universal phenomenon led Einstein to formulate the equivalence principle, according to which a gravitational field is locally equivalent to a field of acceleration. To obtain this principle, he drew upon a fundamental property of gravitational fields already brought to light by Galileo and included in Newton's equations: the acceleration communicated to a body by a gravitational field is independent of its mass.

After the development of special relativity, the need to generalize the theory seemed inevitable for multiple reasons. Relativist unification was far from complete. If the mechanics of free particles and electrodynamics finally satisfied the same laws, it was not the case for Newton's theory of universal gravitation, otherwise the top showpiece of classical physics. Newton's equations are invariant under the classical transformation of Galileo, but not under those of Lorentz. Thus, physics remained split in two, in contradiction with the principle of relativity, which necessitates the validity of the same fundamental laws in all situations.

Moreover, the Newtonian theory is based on certain presuppositions in contradiction with the principle of relativity: it is so with the concept of Newtonian force, which acts at a distance by propagating instantaneously at an infinite speed. Thus, the construction of a relativist theory of gravitation seemed to Einstein (and other physicists) a logical necessity.

Another problem was just as serious: the relativist approach explicitly gives itself the problem of changes in reference systems and their influence on the form of physical laws. But the answer provided by special relativity is only partial. It only considers frames of reference in uniform translation, at constant speeds concerning one another. However, the real world constantly shows us rotations and accelerations, from the multiple forces at work (such as gravity), or inversely, causes new forces (such as the force of inertia).

What are the laws of transformation in the case of accelerated frames of reference? Why would such frames of reference not be as valid for writing physics laws as inertial frames of reference? The answer is that such a question requires a generalization of special relativity.

The originality of Einstein's approach had been, in particular, to bring together two problems, that of constructing a relativist theory of gravitation and that of generalizing relativity to non-inertial systems, into a single endeavor. The equivalence principle made this unity of approach possible: if the field of acceleration and gravitational field are locally indistinguishable, the two problems of describing changes in the coordinate systems, including those which are accelerated and those which are subject to a gravitational field, boil down to a single problem. But such an approach is not reducible to "making relativist" Newtonian gravitation. While certain physicists could hope, at the time, that the problem of Newton's theory could be solved by a simple reformulation, by introducing a force that propagated at the speed of light, it is the entire framework of classical physics that Einstein proposed to reconstruct with general relativity. Better yet, it was a new type of theory which he developed for the first time: a theory of a framework (curved spacetime, now a dynamic variable) in connection with its contents, and no longer only a theory of "objects" in a rigid preexisting framework (as was Newton's absolute space).

Why such a radical choice? Doubtless, because special relativity itself was unsatisfactory on at least one essential point: the spacetime which characterizes it, even if it includes in its description space and time which is no longer absolute taken individually, remains absolute when taken as a four-dimensional "object." However, inspired in particular by the ideas of Ernst Mach, Einstein had come to think that an absolute spacetime could have no physical meaning, but rather, that its geometry should be in correspondence with its material and energetic contents. Thus, a reflection on the problem of inertial forces, which had caused Newton to introduce absolute space, led Einstein to the opposite conclusion.

Relativity of gravity

This was one of the great ideas proposed by Einstein in 1907. This theory was really important as it helped shape the understanding of gravity's basic concept in a manner different from Newtonian mechanics. If an observer descends in free fall within a gravitational field, they no longer feel their weight, which means that they no longer feel the existence of this field itself. At the onset, this idea was viewed with skepticism, but now we have seen how astronauts float weightlessly in their ships, and how objects leave them floating at a constant speed. This idea was revolutionary because it proved that gravity doesn't just exist on its own and the existence of gravity is dependent on the choice of a frame of reference.

Adopting this proposition, he was able to create a distinction from the former idea of gravity. In the Newtonian model, gravity was absolute. It has been recognized as Universal law. It was indeed a physical phenomenon of which the existence does not seem to depend on such a condition of observation.

However, when an enclosed area is allowed to fall freely under gravity, and then we put in motion a body at a certain velocity to this area, the body will move in a straight line at a constant speed about the enclosure walls. In respect to the wall, a body initially immobile will stay like that during the movement of the enclosure's fall. In other words, all experiments that we can perform there would confirm that we are in an initial frame of reference! This meant that although gravity is universal it can be canceled out solely by a judicious choice of the coordinate system. Einstein was able to understand that the understanding of gravity is dependent on the choice of a coordinate system.

Relativity and Quantum Mechanics

The modern world of physics is remarkably based on two main theories, general relativity and quantum mechanics, although both theories use seemingly incompatible principles. The postulates that define Einstein's theory of relativity and quantum theory are supported by rigorous and repeated empirical evidence. However, both resist being incorporated into the same coherent model. Since the mid-twentieth century, relativistic quantum theories of the electromagnetic field (quantum electrodynamics) and nuclear forces (electro devil model, quantum chromodynamics) appeared, but there is no relativistic quantum theory of the gravitational field that is fully consistent and valid for intense gravitational fields (There are approximations in asymptotically flat spaces). All consistent relativistic quantum theories use the methods of quantum field theory.

In its ordinary form, quantum theory abandons some of the basic assumptions of the theory of relativity, such as the principle of locality used in the relativistic description of causality. Einstein himself had considered the violation of the principle of the locality to which quantum mechanics seemed to be absurd. Einstein's position was to postulate that quantum mechanics, although consistent, was incomplete. To justify his argument and his rejection of the lack of locality and the lack of determinism, Einstein and several of his collaborators postulated the so-called Einstein-Podolsky-Rosen (EPR) paradox, which demonstrates that measuring the state of a particle can instantaneously change the status of your linked partner, although the two particles can be at an arbitrarily large distance. Modernly, the paradoxical result of the EPR paradox is known to be a perfectly consistent consequence of the so-called quantum entanglement (that we have seen in chapter 4 and 6).

Chapter 10: Quantum Theory

The development of the quantum theory was not finished with quantum mechanics. Because it has its limits: it doesn't contain the special theory of relativity. Therefore, it only applies to objects that move much slower than the speed of light. But especially the photons, which always move at the speed of light, are not covered by the Schrödinger equation at all. As already mentioned, the English physicist Paul Dirac has not only developed the abstract version of quantum mechanics. In 1928 he succeeded in integrating the special theory of relativity into the Schrödinger equation. This is the Dirac equation, which is an impressive example of the fact that we discover mathematics and thus the physics laws, and not invent them.

But even with the Dirac equation, the photons could not be described. Moreover, like the Schrödinger equation, it only applies to a constant number of particles. The special theory of relativity, however, makes its creation and destruction possible. Both the photons and the variable particle number required an entirely new concept, that's quantum field theory (QFT). However, the step from quantum mechanics to QFT was only a small step in comparison to the enormous revolution from classical physics to quantum mechanics.

Their development began already in the '20s, parallel to quantum mechanics. Paul Dirac and Werner Heisenberg were significantly involved. You already know them. Also, the Italian physicist Enrico Fermi (1901 - 1954) and the Austrian physicist Wolfgang Pauli (1900 - 1958) made contributions. But the development quickly came to a standstill because there were absurdities in the form of infinitely large intermediate values that could not be eliminated for a long time. It was not until 1946 that people learned to deal with them. The first quantum field theory emerged around 1950; it was quantum electrodynamics (QED), the quantum version of the Maxwell equations. The American physicist Richard Feynman (1918 - 1988), a charismatic personality, played a decisive role in its development. He tended to bizarre actions, for example, regularly drummed in a nightclub. And he was involved in solving the Challenger disaster in 1986. Richard Feynman is the only physicist who still gave lectures for beginners when he was already famous. That was in the early '60s. They have also been published in book form and are still widely used today. The German textbooks on physics are, as expected, factual and dry. The Feynman textbooks are more relaxed and contain much more text. My impression: For the first contact with physics, they are less suitable, but they improve their understanding if one has already learned some physics.

QED is the foundation of everything that surrounds us. The entire chemistry and thus also biology follow from it. But even with it, the development of the quantum theory was not yet complete. Because two more forces were discovered, the strong and the weak force. They only play a role in the atomic nuclei, so we don't notice them. We only notice the gravitational and the electromagnetic force. Even though large bodies are always electrically neutral, QED also plays an important role in everyday life. The matter is almost empty. Accordingly, if two vehicles collide, they should penetrate each other. If it wasn't for QED.

The strong and weak force led to the development of two further QFTs. Whereby the QFT of the weak force could be combined with the QED. For the sake of completeness, the QFT of the strong force is called quantum chromodynamics (QCD). The QFTs of the three forces that can be described with them are combined to form the so-called standard model of particle physics. It represents the basis of modern physics.

In the QFTs, each type of elementary particle is described as a field in which particles or quanta are created and destroyed. This is the core of the QFTs. What are the elementary particles? Matter consists of molecules that consist of atoms that can be separated into nucleus and electrons. The electrons are called elementary particles because they cannot be further divided according to today's knowledge. And there is another elementary particle, the Higgs particle, which was discovered at the Large Hadron Collider (LHC) in Geneva in 2012. It plays a special role; I won't go into it further.

Since the general theory of relativity cannot be integrated into the standard model of particle physics, there must be a more fundamental theory. String theory was a promising candidate, but the hope is fading.

Will the more fundamental theory, if it is ever found, be the "theory of everything"? So will it herald the end of theoretical physics? Some believe this, such as the recently deceased physicist Stephen Hawking (1942 - 2018). So far, however, every physical theory always has something difficult in it. One can hope that we will find a theory that explains everything. But I doubt that we will ever succeed. Therefore, even in the far future, there should be theoretical physicists who are looking for new theories.

In the 17th century, Sir Isaac Newton united Galileo Galilei's and Johannes Kepler's research on the acceleration and planetary motion in his principal Principia Mathematica. The gravitational law was born. The well-known three fundamental laws of the movement also arose with this work. Newton not only laid the foundations for mechanics but also created the classic Newtonian world image, that many things could finally be explained logically. Apples falling from the trees or planets around the sun were no longer a mystery, but facts that could be adequately described by Newton's laws. From then on, the world was ticking like complex clockwork that was deterministic and causal. Everything was descriptive and calculable. However, two hundred years later and at the beginning of the twentieth century, a German physicist's discovery should completely shatter the Newtonian world.

Max Planck the Father of Quantum Theory

All objects emit electromagnetic radiation, which is called heat radiation. But we only see them when the objects are very hot. Because then they also emit visible light. Like glowing iron or our sun. Of course, physicists were looking for a formula that would correctly describe the emission of electromagnetic radiation. But it just didn't work out. Then, in 1900, the German physicist Max Planck (1858 - 1947) took a courageous step.

The emission of electromagnetic radiation means the emission of energy. According to the Maxwell equations, this energy release should take place continuously. "Continuously" means that any value is possible for the energy output. Max Planck now assumed that the energy output could only take place in multiples of energy packets, i.e., in steps. That led him to the correct formula. To the energy packets, Planck said, "quanta." Therefore, the year 1900 is regarded as the year of birth of the Quantum Theory.

In 1905 an outsider named Albert Einstein was much more courageous. He took a closer look at the photoelectric effect. It means that electrons can be knocked out of metals by irradiation with light. According to classical physics, the electrons' energy knocked out should depend on the intensity of the light. Strangely enough, this is not the case. The energy of the electrons does not depend on the intensity but the frequency of the light. Einstein could explain that. For this back again to the quanta of Max Planck. The energy of each quantum depends on the frequency of the electromagnetic radiation. The higher the frequency, the greater the energy of the quantum. Einstein now assumed, in contrast to Planck, that the electromagnetic radiation itself consists of quanta. It is the interaction of a single quantum with a single electron on the metal surface that causes this electron to be knocked out. The quantum releases its energy to the electron. Therefore, the energy of the electrons knocked out depends on the frequency of the incident light.

However, the skepticism was great at first. Because electromagnetic radiation would then have both a wave and a particle character. But as we have seen, another experiment also showed its particle character. This was the experiment with X-rays and electrons carried out by the American physicist Arthur Compton (1892 - 1962) in 1923. As already mentioned, X-rays are also electromagnetic radiation, but they have a much higher frequency than visible light. Therefore, the quanta of X-rays are very energetic. That's why they can invade the human body. But that makes them so dangerous. Compton was able to show that X-rays and electrons behave similarly to billiard balls when they meet.

What are photons? That is still unclear today. Under no circumstances should they be imagined as small spheres moving forward at the speed of light. Because the photons are not located in space, so they are never at a certain place. Here is a citation from Albert Einstein. Although it dates back to 1951, it also applies to today's situation: "Fifty years of hard thinking have not brought me any closer to the answer to the question "What are light quanta?" Today, every Tom, Dick, and Harry are imagining they know. But they're wrong."

How Max Planck Developed the New Concepts of Physics

All objects in the universe are constantly exchanging energy in the form of electromagnetic radiation. Each object emits, partially absorbs, and partially reflects electromagnetic waves. The amount of the object's energy is directly evidenced by its temperature, i.e., the energy of its atoms' vibrations (thermal motion).

Most of the objects around us are of regular temperature, so their emissions are weak and almost exclusively in the infrared range. However, the higher the temperature, the more energy they have, and the more they emit this energy, the more the objects cool off. First, the visible light is added to the infrared light with the largest wavelength, about 750nm, i.e., the red light adjacent to the infrared range in the spectrum. As the temperature rises, the light with the shorter and shorter wavelength is gradually added (colors following across the breadth of the visible light spectrum), until a sufficiently high temperature brings up the whole spectrum of colors, resulting in whiteness.

Moreover, the ultraviolet (UV) range is also gradually added after the visible light: more and more, extremely shorter waves. Thus, as the object's temperature rises, the radiation range gradually expands from the starting infrared range towards UV frequencies. But why is the radiation changing that way? Why do the objects not radiate the entire radiation range evenly?

In fact, according to the laws of classical physics known at the end of the 19th century, the farther from the infrared radiation in the UV direction of the spectrum (including the adjacent X-rays, then gamma rays), the more energy should be emitted so that the total energy in the whole range had to be infinite. This divergence between theory and reality was called the ultraviolet catastrophe.

Now, as objects never have an infinite quantity of energy, it would make the most sense if they just emitted all the present energy at once. In practice, that would look like the world is full of radiation, gamma radiation mostly, and much colder objects, which would be very hard to heat.

As already mentioned, in addition to emitting their energy, objects also reflect a certain amount of radiation from surrounding objects. This complicates the overall picture if you are precisely trying to measure only the emitted radiation, which must somehow be distinguished from the reflected radiation.

Therefore, to simplify the research, physicists considered a model of a body that does not reflect rays but only absorbs and emits, which was known as the black body, as it is dark and can never be seen clearly because any light that falls on it is absorbed, rather than being reflected into our eyes, thereby making the object visible.

As is usually the case in practical science, scientists have never had a perfect black body. Still, some inventions reflected radiation very little and allowed scientists to conduct experiments much more precisely than with usual objects. This would guarantee that they were dealing with the emitted radiation that depends on the object's temperature.

Scientists were looking for some additional explanation of why the radiation had a limited range and quantity. It was one of the greatest scientific mysteries of the time, especially since the established laws of physics in the 19th century – which said that objects must give off their energy with unlimited intensity – worked perfectly on other important issues. As it turned out, the solution was so unusual and revolutionary that for years it wasn't taken seriously by the general physicists' community or even the author of the solution himself!

As everybody knows, we can use a fractional number to express any value, such as length, weight, or energy, to any degree of accuracy. For instance, if the whole number 4 is not precise enough, we can use a fraction to specify the value to greater extents, for example, 4.5, or 4.531, or even 4.531112191, and so on.

There is also no such thing as a minimum number to describe a value since numbers can be fractionally smaller and smaller without end: 0.01 => 0.001 => 0.0001, etc. These very ideas about any energy amount defined physics until the 20th century, and they had to be modified according to the original mathematical model of radiation depending on the temperature created by Max Planck (known as Planck's law).

Planck did not claim his model provided a comprehensive explanation of the limited radiation of heated objects. Planck's law was intended only as a convenient temporal mathematical model to reflect the observed radiation distribution to a fair degree. It was created to enable more-or-less adequate practical calculations for radiation at high temperatures (in particular, to predict the radiation energy and color of light of a tungsten filament depending on temperature). In simple words, Planck's law has the following logic: since the shorter the wave, the more limited the amount of radiation, Plank suggested imagining that the radiation always consisted of single indivisible portions or pieces of energy – quanta (just as any substance consists of individual atoms or molecules), as well as imagining that the radiation of the shorter wavelength is composed of larger (meaning, more energy) portions.

Therefore, more radiation of greater wavelength would be produced, since it is easier for energy scattered in bodies to gather in smaller portions of radiation of greater wavelengths, than in the larger portions of short-wave radiation. As objects are heated to higher temperatures, a significantly larger amount and density of energy enables the formation of bigger and bigger portions of shorter and shorter wavelength radiation, i.e., the amount of radiation of shorter wavelengths gradually increases, which is precisely how it is observed in reality.

This also implies that the lowest possible value of radiation power is one quantum (bigger or smaller depending on the wavelength). By this model, radiation cannot contain ½ a quantum of energy, nor can it contain 1½ or 109¼ quanta. That is, quanta are only measurable in whole numbers, not fractions. As mentioned before, an object's temperature is the energy of the motion of the atoms forming the object.

If the temperature is low, this means that the atoms have a low energy level. If the energy level is low, then each atom or each certain group of atoms will produce, on average, smaller quanta (i.e., radiation of a longer wave) than when the energy level in atoms is higher.

And if the amount of energy in an atom or a group of atoms is not enough to form one quantum of UV (or X-rays or gamma rays), then this radiation can never be emitted. Ordinary-temperature objects do not have enough energy to emit even middling-energy quanta of light, but only lower-energy quanta of low frequency (long wavelength) microwave, infrared, and radio waves.

The greatest peculiarity of Planck's model lies in the fact that, for some reason, energy is emitted and absorbed only in individual quantities. Any radiation dose is measured in a certain specific number of these smallest radiation units. Just as any amount of gold is a certain number of gold atoms, and just as molecules and atoms have different sizes, radiation consists of quanta (particles of energy!) of a size-specific only to that particular wavelength (or corresponding frequency).

Summing up, Planck suggested that the shorter the wave (or the higher the wave frequency), the larger the quantum, thus reducing the radiation of shorter waves (higher frequencies). He found that the amount of energy per radiation quantum must always be equally proportional to that radiation frequency. To find out the amount of the energy of the radiation quantum of any particular frequency, multiply this particular frequency by the standard coefficient (the Planck constant h).

Now Planck's formulas allowed the describing and predicting of the radiation of heated objects precisely rather than approximately! Splitting the quantities that seemed to be continuous into quanta solved some other physics problems as well. This splitting of quantities into elementary minimal indivisible units is called quantization. The quantization means the quantities can gain not just any value, but only the allowed whole values, like 1 or 109 or 1,000,000 quanta of energy. It will be shown that this idea also works when considering the structure of an atom.

The general effectiveness of this simple concept of quanta makes it virtually indisputable. Quantum physics, which began with this concept, is the most successful, most productive, and precise theory created by humankind. Its unprecedented benefits include all of the forms of modern, sophisticated technology (thanks to the understanding of semiconductors and the building of transistors with them), and its constant, unremitting development promises further new opportunities that can hardly be imagined today. High-tech experiments used to test the rest of the unusual concepts of quantum physics just prove it even further.

However, in the 1900s, after Planck suggested his convenient model, still no one doubted that the amount of radiation energy could be of any value (and not of a certain number of quanta). Why should energy be split into elementary indivisible units? It seemed contrived and not believable.

For five years afterward, everybody, along with Planck believed that he had managed to invent only a temporary artificial schematic approach to the issue of radiation, which worked thanks to a complete fluke, and a real understanding of the observed radiation behavior was yet to come.

However, the future held another powerful argument in favor of Planck's concept of quanta. Albert Einstein realized that the second great scientific mystery of that time, the photoelectric effect, could be easily explained by quanta's existence. When radiation is absorbed by objects, its energy can be added to the motion energy of its atoms or molecules, causing the object's temperature to rise.

On the other hand, radiation energy can be transmitted to individual electrons in atoms. If there is enough energy, an electron can overcome the attraction of the nucleus. Thus, electrons leave their atoms upon the impact of radiation, such as light (photoelectric effect).

It was known that even a very large amount of long-wave red light (about 650-700nm) could not liberate electrons from atomic binding. In contrast, even the relatively small power of violet light (about 400nm) and ultraviolet radiation (400 to 10nm) could liberate electrons effectively, even if the total energy of violet light or UV radiation is hundreds of times lower than red light energy.

So, the total radiation power was not the issue. Violet light and UV radiation had to have some other advantage, although it was logical to assume that to liberate electrons from atomic binding, it was necessary to apply as much energy as possible to these electrons.

Einstein pointed out that violet light and UV radiation have a great advantage over the red light in Planck's model because they had a greater power of individual "units" of radiation – Planck's quanta – than red light.

Einstein also had to assume that to break away from the atom, the electron, rather than gradually accumulating the energy of individual small quanta of light or absorbing many small quanta simultaneously, was capable of absorbing, for some reason, no more than a single quantum with enough energy to allow the electron to instantly overcome the stretching of the nucleus. Such a bold step in the understanding of quantum reality, second after Planck, brought Einstein his Nobel Prize.

The Bohr atomic model

In 1897, the British physicist Joseph John Thomson (1856 - 1940) discovered electrons as a component of atoms and developed the first atomic model, the so-called raisin cake model. Therefore, the atoms consist of an evenly distributed positively charged mass in which the negatively charged electrons are embedded like raisins in a cake batter. This was falsified in 1910 by the New Zealand physicist Ernest Rutherford (1871 - 1937). With his experiments at the University of Manchester, he was able to show that the atoms are almost empty. Another form of movement was inconceivable at that time. That led physics into a deep crisis. Because the electrons have an electrical charge, and a circular motion causes them to release energy in the form of electromagnetic radiation. Therefore, the electrons should fall into the nucleus. Hence the deep crisis, because there should be no atoms at all.

In 1913 a young colleague of Ernest Rutherford, the Danish physicist Niels Bohr (we have "met" both of them multiple times in the previous chapters), tried to explain the stability of atoms. He transferred the idea of quanta to the orbits of electrons in atoms. Without, however, being able to explain why this should be the case. Nevertheless, his atomic model was initially quite successful because it could explain the so-called Balmer formula. It has been known for some time that atoms only absorb light at certain frequencies. They are called spectral lines. In 1885, the Swiss mathematician and physicist Johann Jakob Balmer (1825 - 1898) found a formula with which the spectral lines' frequencies could be described correctly. But he couldn't explain them. Bohr then succeeded with his atomic model, at least for the hydrogen atom. This is because electrons can be excited by photons, which causes them to jump on orbits with higher energy. This is the famous quantum leap, the smallest possible leap ever.

Since only certain orbits are allowed in the Bohr atomic model, the energy and thus the frequency of the exciting photons must correspond exactly to the energy difference between the initial orbit and the excited orbit. This explained the Balmer formula. But Bohr's atomic model quickly reached its limits because it only worked for the hydrogen atom. The German physicist Arnold Sommerfeld (1868 - 1951) expanded it, but it still represented a rather unconvincing mixture of classical physics and quantum aspects. Besides, it still could not explain why certain orbits of the electrons should be stable. Sommerfeld had a young assistant, Werner Heisenberg (1901 - 1976), who, in his doctoral thesis, dealt with the Bohr atom model extended by Sommerfeld. Of course, he wanted to improve it. In 1924 Heisenberg became assistant to Max Born (1882 - 1970) in Göttingen. The breakthrough came a short time later, in 1925, on the island of Helgoland, where he cured his hay fever. He was able to explain the frequencies of the spectral lines, including their intensities using so-called matrices. He published his theory in 1925, together with his boss Max Born and Pascual Jordan (1902 - 1980). This is considered to be the first quantum theory and is called matrix mechanics. I will not explain it in more detail because it's not very clear. And because there is an alternative mathematically equivalent to it, which enjoys much greater acceptance because it is easier to handle. It is called wave mechanics and was developed in 1926, just one year after matrix mechanics, by the Austrian physicist Erwin Schrödinger (1887 - 1961).

The Schrödinger Equation

Erwin Schrödinger in 1926 introduced the equation named after him. The circumstances surrounding its discovery are unusual. It is said that Schrödinger discovered it at the end of 1925 in Arosa, where he was with his lover.

The Schrödinger equation is at the center of wave mechanics. As already stated, it is mathematically equivalent to Heisenberg's matrix mechanics. But it is preferred because it is much more user-friendly. There is a third version, more abstract, developed by the English physicist Paul Dirac. All three versions together form the non-relativistic quantum theory called quantum mechanics. As you rightly suspect, there is also a relativistic version.

The Schrödinger equation is not an ordinary wave equation, as it is used, for example, to describe water or sound waves. But mathematically, it is very similar to a "real" wave equation. Schrödinger could not explain why it is not identical. He had developed it more out of intuition. According to the motto "What could a wave equation for electrons look like?" This can also be called creativity. Very often, in the history of quantum theory, there was no rigorous derivation. It was more of a trial and error until the equations that produced the desired result were found. Strangely, a theory of such precision could emerge from this. However, as I will explain in detail, the theory is also mastered by problems that have not yet been solved.

The solutions of the Schrödinger equation are the so-called wave functions. It was only with them that the stability of the atoms could be convincingly explained. Let us consider the simplest atom, the hydrogen atom (we have described it in Chapter 7). It consists of a proton as the nucleus and an electron moving around it. If one solves the Schrödinger equation for the hydrogen atom, then one finds that the electron assumes only specific energy values. There must be the smallest energy value. This means that the electron always has a certain distance from the nucleus, which makes the hydrogen atom stable. Bohr's atomic model also provides specific energy values but cannot justify them. Does that make the Schrödinger equation better? Not really, it is the used mathematical formalism that led to the specific energy values. Why it is precisely this formalism that fits reality, nobody can answer until today.

Chapter 11: Elastic Shocks and Inelastic Shocks

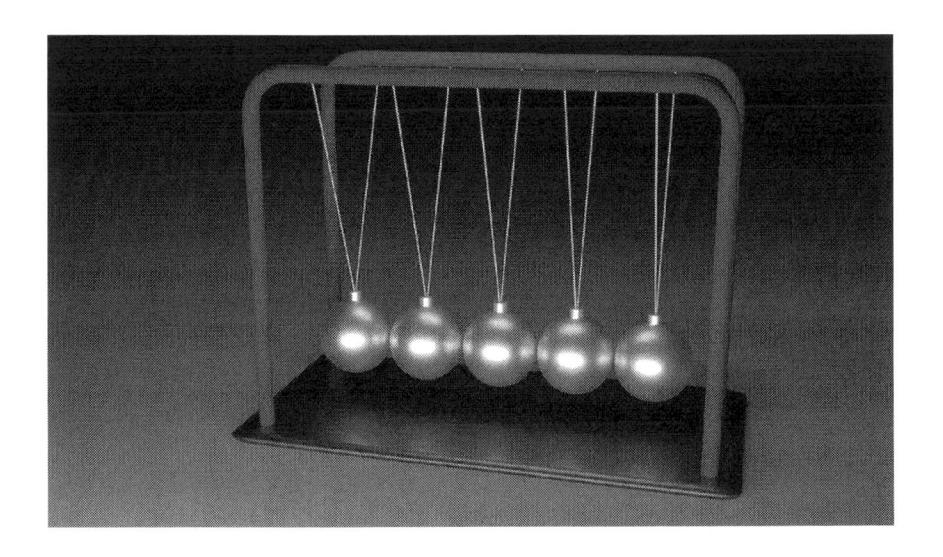

Metals and Insulators

Those of us lucky enough to encounter it realize how fundamental power is to present-day living. It uncovers the force we use in our homes to prepare our food and drive the PCs that process our information. This section plans to clarify how this is a sign of quantum physics standards and, specifically, how quantum physics permits us to understand why the electrical properties of various solids can change from metals that promptly lead power to insulators that don't. Initially, I should accentuate again that power isn't a wellspring of energy in itself, yet rather a method of transmitting energy starting with one place then onto the next. Power is created in a force station, which thus gets its energy from some type of fuel (for example, oil, gas, or atomic material) or maybe from the breeze, waves, or sun. Figure (a) below shows a power station driving an electric current through one wire to a computer and back along with another. In order to conduct electricity, it is required that the wires are made of metal. Figure (b) shows a simple electric circuit. A current I is driven around the circuit by the Voltage V created by a battery.

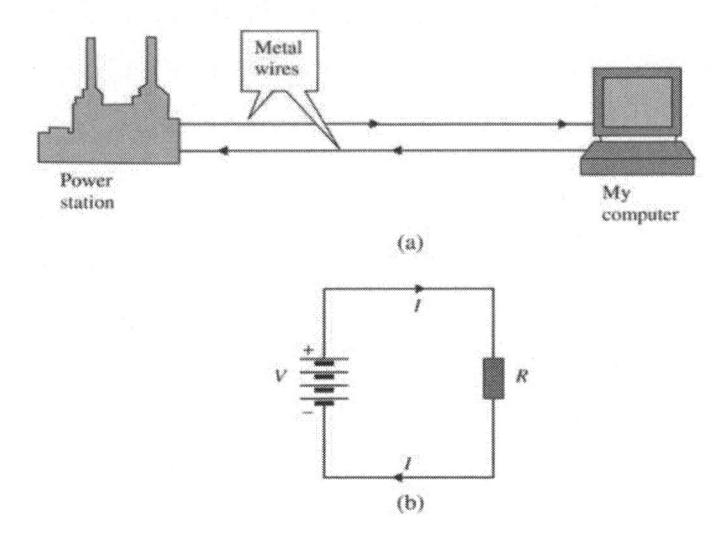

Through the resistor R, because electrons are negatively charged, they move opposite to that of the conventional current.

We have to understand how this occurs and what these terms mean. To begin with, the battery: this comprises various 'electrochemical cells' that use a synthetic procedure to produce positive and negative electrical charges on the furthest edges of every cell. These would then be able to apply power on any versatile charges associated with them, and their potential for doing so is named the 'voltage' created by the battery.

Next, the connecting wires: these are made of metal, and metals are materials that contain electrons that can move openly inside the material. When a wire is associated with a battery as electrons near the negative terminal of the battery experience an appalling power driving them through the wire. They travel around the circuit until they arrive at the positive terminal to which they are pulled in; they then go through the battery and develop at the negative terminal where the procedure is repeated. Accordingly, there is a current stream around the circuit and we should note that because the electrons convey a negative charge. The regular course of the current stream is inverse to that of the electrons. The purpose behind this is just that the idea of electric flow was created, and the ordinary significance of positive and negative charge was set up before electrons were found. The voltage expected to drive a given current through a given resistor is corresponding to the size of the current and the obstruction; this relationship is known as 'Ohm's law', which we will talk about in more detail towards the end of this part. Resistors are frequently developed from specific metal composites, intended to introduce critical protection from the progression of current; the current that moves through them loses a portion of its energy, which is changed over into heat. This is the procedure that underlies any electrical warmer activity. For example, it might be found in a pot or a clothes washer or used to warm a room.

In this image, we can see an electric circuit formed by a loop of wire. A whole number of wavelengths of the electron waves must match the distance around the circle.

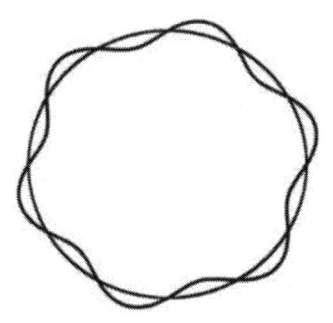

A few materials, for example, glass, wood, and most plastics are 'insulators' whose opposition is high to such an extent that they don't permit the entry of any electrical flow. They assume an essential job in the electrical circuits planning since they can be used to isolate flow conveying wires, thus guaranteeing that electrical flows stream where we need them. Readers are likely acquainted with this in a household setting, where the wires hefting electrical flow around our homes are ensured by plastic sheaths that forestall them connecting or coming into contact with ourselves.

The contrast among metals and insulators is one of the most emotional of any known physical properties. A good metal can lead power well over a trillion times more proficiently than is a good insulator. However, we realize that all materials are made out of particles that contain electrons, protons and neutrons. How is it that their properties can be so extraordinary? Generally, we will see that the appropriate response lies in quantum physics: if the electrons were not quantum objects with wave properties, none of this would be conceivable. A strong metal is somehow like a goliath particle: the molecules are held very near one another, and the external electrons are not, at this point, bound to specific atoms. Starting now and into the foreseeable future, we will recognize these 'free' electrons and the decidedly charged 'particles', by which we mean the core of a molecule alongside its inward shell electrons. As a first estimation, we will accept that the free electrons' behavior is unaffected by the particles. Later we will perceive how the nearness of the particles must adjust this image, and we will find that if the particles structure a customary exhibit, as they do in a precious stone, the wave properties of the electrons imply that their movement is to a great extent unhindered in metals, yet completely discouraged in insulators.

No resistance

When resistance disappears? The answer to this question was given by Kamerlingh Onnes as early as 1914. He proposed a very ingenious method of measuring the resistance. The experimental scheme looks easy.

He had lead wires from the coil in a cryostat, an apparatus for carrying out experiments at low temperatures. The cooled helium coil is superconducting. In this case, the current flows through the coil, creating a magnetic field around it, which can be easily detected by the deviation of the magnetic needle located outside the cryostat. Then the next step was to close the key, so that now the superconducting stroke is short-circuited. However, a compass needle was diverted, indicating that the presence of current in the coil was already disconnected from a current source. Watching the arrow for a few hours (until evaporating the helium from the vessel), Onnes had not noticed the slightest change in the deflection direction.

According to the results of the experiment, Onnes concluded that the resistance of the superconducting lead wire is at least 10^{11} times lower than its resistance in the normal state. Subsequently conducting similar experiments, it was found that the current decay time exceeds many years, and this indicated that the superconductor resistivity is less than 10^{25} ohms x m. Comparing this with a resistivity of copper at room temperature, the difference is so large that one can safely assume that the resistance of the superconductor is zero. It is difficult to call another monitor and modify the physical quantity which would be addressed in the same "round zero" as the conductor resistance below the critical temperature.

Recall from school physics course Joule - Lenz: the current I flowing through the conductor with a resistance R generates heat. Heating limits the throughput power, for example the power of electric cars. Thus, in particular, is the case with electromagnets. Obtaining strong magnetic fields requires a large current, which leads to the release of an enormous amount of heat in the solenoid's windings. But the superconducting circuit remains cold, the current will circulate without damping - impedance equal to zero, no power losses.

Since the electric resistance is zero, the exciting current of a superconducting ring will exist indefinitely. In this case, the electric current resembles the current produced by the electron orbit in the Bohr atom: it is like a very large Bohr orbit. Persistent current and the magnetic field generated by it cannot have an arbitrary value, they are quantized so that the magnetic flux penetrating the ring takes values that are multiples of the elementary flux quantum F on = h / (2e) = 2.07 10^{15} Wb (h - Planck's constant).

Unlike electrons in atoms and other microparticles, whose behavior is described by quantum theory, superconductivity is a macroscopic quantum phenomenon. Indeed, the superconducting wire's length through which flows persistent current can reach many meters or even kilometers. Thus, carriers describe a single wave function. This is not the only macroscopic quantum phenomena. Another example is superfluid liquid helium or substance neutron stars.

In 1913 Kamerlingh Onnes was proposing to build powerful electromagnet coils of superconducting material. Such a magnet would not consume electricity, and it could receive a super-strong magnetic field.

If so...once tried to pass through the significant superconductor current, the superconductivity disappears. Soon it turned out that a weak magnetic field also destroys superconductivity. The existence of critical values of temperature, current and magnetic induction severely limits superconductors' practical possibilities.

Electrical resistance superconductors

The most accurate method of measuring low impedances consists of measuring the current decay time induced in the closed circuit of the test material. The decrease in current time energy LI 2 /2 (L - inductance loop rate) consumed for Joule heat: here integrating (I 0 - current value at t = 0, R - resistance of the circuit).

The current decays exponentially with time, and the attenuation rate (at a given L) is determined by the electrical resistance.

For small R, a formula can be written as dI-current change during Δt. Experiments conducted using a thin-walled superconducting cylinder with extremely small values of L showed that the superconducting current is constant (with accuracy) within a few years. It followed that the resistivity in the superconducting state is less than 4 x 10^{25} ohms or more than 10^{17} times less than the resistance of copper at room temperature. Since the possible decay time is comparable to the time the existence of mankind, we can assume that R DC in the superconducting state is zero.

Thus, the superconducting current is only like a real-life example of perpetual motion on the macroscopic scale!

When R = 0, the potential difference V = IR on any segment of the superconductor and hence the electric field E inside the superconductor is zero. The electrons create a current in the superconductor, moving at a constant speed without being scattered by the thermal vibrations of the crystal lattice atoms and its irregularities. Note that if E is not equal to zero, electrons carrying the superconducting current are accelerated without limit and the current could reach an infinitely large value, which is physically impossible. To create a superconducting current, it is necessary only to accelerate the electrons up to a certain speed directional movement (expending this energy). Further, the current is constant without borrowing power from an external source (as opposed to a conventional current conductor).

The situation changes if the superconductor is applied to the variable potential difference as it creates a variable superconducting current. During each period, the current changes direction. Consequently, in the superconductor must exist an electric field that periodically slows down the superconducting electrons and accelerates them in the opposite direction. Since it consumes energy from an external source, the electrical resistance of alternating current in a superconducting state is zero. However, because the electron mass is very small, power loss at frequencies less than 10^{10} - 10^{11} Hz is negligible.

Superconductors in a magnetic field

It is certain that when a magnetic field is greater than a threshold or critical value, the superconductivity disappears completely. Even if some metal would lose resistance when cooled, it cannot go back to normal, once an external magnetic field is applied. In this case, too, the metal is recovered at about the resistance that it was at a temperature above the temperature T superconducting transition.

Consider now the behavior of a perfect conductor (i.e., conductor deprived of resistance in different environments). When such a conductor is cooled below a critical temperature, the electrical conductivity becomes infinite. This property has allowed considering a perfect superconductor conductor.

An ideal conductor's magnetic properties emerged from the law of induction of Faraday (you may know him for his famous cage) and conditions of infinite conductivity. Assume that a metal transition to the superconducting state occurs in the absence of a magnetic field, and an external magnetic field is applied only after the disappearance of resistance. It does not need any subtle experiments to ensure that the magnetic field inside the conductor does not penetrate. Indeed, when metal enters the magnetic field, on its surface due to electromagnetic induction the resultant magnetic field is zero.

A situation arises in which the metal prevents the penetration of the magnetic field, which behaves as a diamagnetic. Now, if the external magnetic field is removed, the pattern will not be magnetized.

Now put in the magnetic field of the metal in the normal state, and then cool it to be transferred to the superconducting state. The disappearance of the electric resistance should not influence the magnetization of the sample, and therefore the magnetic flux distribution will not change it. Suppose now the applied magnetic field is removed. In that case, the change in the flux of the external magnetic field through the sample volume will result (according to the law of induction) in the appearance of persistent flows, a magnetic field which exactly compensates for the change in the external magnetic field. As a result, the captured field will not be able to escape; it will be "frozen" in the amount of sample and remain there in a kind of trap.

As can be seen, the magnetic properties of an ideal conductor depend on how it gets to the magnetic field. In fact, at the end of these two operations - the application field and reducing - the metal is in the same conditions - at the same temperature and zero external magnetic fields. However, the magnetic induction-metal sample in both cases is quite different - zero in the first case and the end, depending on the source field in the second.

Tunnel effect

The tunnel effect is known in physics for a long time. This effect was first predicted in the early 20th century, but the acceptance of it as a general physical phenomenon came only mid-century. It is a quantum-mechanical process by which a particle can pass through a potential energy barrier that is higher than the energy of the particle. It was first postulated to explain the escape of alpha particles from atomic nuclei.

A particles (for example, an electron in the metal) approaches a barrier (e.g., a dielectric layer), however based on the classical idea, it would not be able to pass through it, as its kinetic energy is insufficient. On the contrary, according to quantum mechanics, the barrier passage is possible. The particle may have a chance, as it were, to pass through the tunnel through a classically forbidden region. Quantum mechanics for the microparticles (electron) holds uncertainty relation $\Delta h \Delta r > h$ (x - coordinate of the particle, p - its pulse). When a small uncertainty of its coordinates in a dielectric $\Delta h = d$ (d - thickness of the dielectric layer) leads to a significant uncertainty its pulse $Dp \geq h / \Delta x$, and consequently, the kinetic energy p 2 / (2m) (m - the mass of particles), the energy conservation law is not violated. Experience shows that indeed between two metal electrodes separated by a thin insulating layer (tunnel barrier), electric current can flow the greater, the thinner the dielectric layer.

Understanding Quantum tunneling is important to understand the Quantum world, as this effect plays an essential role in many physical phenomena, like the nuclear fusion that occurs in main sequence stars like the Sun. Other important applications are the tunnel diode, the scanning tunneling microscope and quantum computing. Quantum tunneling is one of the newest and most intriguing implications of quantum mechanics, and it carries quite a few implications. For instance, it could pose physical limits to microprocessors transistor's size, due to the ability of the electrons to tunnel past transistors that are too small in size.

You may wonder how this tunnel effect is possible. If you recall, earlier in the book we talked about the Heisenberg uncertainty principle (if you need a refresher, you can go back to Chapter 6). If we try to connect the dots, we should be able understand how this principle can explain the concept of tunneling. We have seen that a quantum object can behave as a wave or as a particle in general. In other words, the uncertainty in the exact location of light particles allows these particles to break rules of classical mechanics and move in space without passing over the potential energy barrier. Isn't that fascinating!

Chapter 12: Practical Applications

Applications of Quantum Theory

Quantum technologies have increased, and today we cannot drive to grandma's house or buy food without taking advantage of quantum physics.

Nevertheless, today's quantum technologies pale in comparison to potential new avenues in the future. The applications of quantum physics to improve health, faster computers, and safer communications are behind the horizon.

The Neon Light

Neon light was first demonstrated in 1855 by the German physicist Heinrich Geissler. He noted that a slight glow was emitted when an electric field was applied through a gas tube containing low-pressure gas.

Nowadays, we know that the applied electric field was stripping the electrons from the atoms in the gas and creating a flow of negative electrons in one direction and positively charged ionized (ions) atoms in the other.

Collisions between fast-flowing ions or electrons with atoms lead to other ionizations, thus continuing the process. This set of electrons and ions is called a plasma. When we think of solids, liquids, and gases as the only three states of matter, physicists consider plasma as a fourth state.

Collisions between ions or electrons and atoms do not always have enough energy to release atomic electrons. When they do not, colliding atoms are only empowered from their ground state to an excited state. Shortly after that, the atom will transition down to a lower-lying state, thus emitting a photon at a frequency set by the spacing between the energy levels.

These photons are responsible for the glow, and the characteristic frequency defines its color. Red light is emitted when a light discharge uses neon, while helium emits a purple color, carbon dioxide emits a white color, and mercury emits a blue color.

This physical process inspired French engineer Georges Claude to formulate a patent for the technology in France in 1910, which led to the use of neon lights for advertising and art. The light discharge is also the basis of the sodium vapor lamp, whose light yellow-orange glow is used to illuminate many streets around the world.

A miniaturized device that operates on the same principles, the neon incandescent lamp, was introduced in 1917. These were used in the

1970s for electronic displays and today serve as the necessary technology for televisions and plasma displays.

Even ordinary everyday fluorescent bulbs are based on light discharge. Here, the exhaust emissions come from mercury vapor, and these are in the (invisible) range of ultraviolet.

The Laser

The laser is one of the best examples of quantum application because it is widely used.

We have already seen that excited atoms emit photons by making a quantum leap to a lower energy state. In most cases, this occurs without any external influence, and the emissions of this variety are called spontaneous.

This is only half the story since atoms can also be driven to emit photons with a process known as stimulated emission.

Stimulated emission is an essential physical process first envisaged by Einstein in 1917.

Before it became so used, "laser" was an acronym for "light amplification through stimulated emission of radiation".

The first successful laser was developed by the American physicist Theodore Maiman in 1958. He found a similar effect, "microwave amplification by stimulated radiation emission". The short name given to this was the maser, and its realization was awarded the Nobel Prize in 1964.

The stimulated emission of a photon occurs when a photon of the same frequency hits an excited atom, that is, a photon stimulates the emission of an equal photon.

Furthermore, the stimulated photon not only has the same frequency as the incident photon but also emerges in the same direction and phase.

The stimulated emission requires an incoming photon and an excited atom. The excited state's energy must be $\Delta E = hf$ above the ground state (h is Planck's constant, f-frequency, and E-energy).

Otherwise, the atom cannot make a quantum leap and emit a photon at the same frequency.

Now imagine that you have an extensive collection of atoms and that many of them are excited $\Delta E = hf$ above the earth state. If you introduce a photon at frequency f, it can stimulate a quantum jump in one of the excited atoms to end up with two equal photons. Each of these two photons can then go on and stimulate two other identical photons, resulting in four "cloned" photons. If you have a vast population of excited atoms, you will have a large army of equal and cloned photons due to the waterfall effect.

This is precisely the "amplification" of the stimulated emission from which the laser derives its name.

Our description of amplification is based on having many excited atoms.

However, atoms prefer to be in their ground state. So, something needs to be done to prepare an extensive collection of excited-state atoms. This is called population reversal, and there are many ways to achieve it. In all cases, however, it is necessary to have more atoms in the excited state than in the earth state. Otherwise, more photons will be absorbed than those emitted, and the waterfall will run out.

The process can be helped by placing a pair of mirrors on each end of the collection of atoms. One mirror should be 100% reflective, while the other should be partially transparent. This will allow a coherent, one-way beam of light to escape from one end and continue doing useful things.

The first laser in the world consisted of a flash lamp coiled around a ruby rod, inserted between a pair of mirrors. The flashing lamp generated a population reversal of atoms in the ruby crystal and mirrors bounced photons back and forth to build a cascade of stimulated photons.

Today, lasers are available in all shapes and sizes.

The population of atoms, or "laser medium" is usually some form of gas or solid. Also, different lasers use different population inversion schemes. Lasers can also be set to emit light continuously or in pulses, and the energy of the pulse and the duration of the pulse can vary widely.

You meet lasers of many varieties every day.

For example, in the supermarket checkout lane, lasers are used to understand the price of the purchased items by scanning the laser beam through the bar code. The item's price is obtained by a detector that measures the laser light reflected by the barcode.

CD and DVD players scan the discs' surface, where small pits have been burned to digitally encode the images and music that we are trying to see and hear.

Besides, lasers are also used by laser printers, which use them to transfer toner to printed pages.

Lasers have also revolutionized medicine.

The small pulsed lasers are used in surgery because they can emit energy in a very precise way and in tiny places, thus preventing unnecessary damage to nearby organs or tissues. They are also handy and minimally invasive in eye surgery, including retinal reattachment, vision correction.

The GPS

Maybe you have already heard of the Global Positioning System – or indeed its acronym, GPS – since most probably you have a navigator or cell phone.

Nevertheless, do you know that GPS would never have been born except for the laws of quantum physics? This is because, on every GPS satellite, there is an atomic clock.

From alarm clocks to Swatch wristwatches, almost every watch counts time by recording something that occurs at a specific regular frequency.

The pendulum of a pendulum clock, for example, swings back and forth about once per second (or a frequency of one hertz), so about 60 of these equals one minute.

A modern wristwatch is based on a quartz crystal, which oscillates more than 10,000 times per second. It takes many more cycles to count one minute, but the principle is the same.

We can also measure the frequency using a quantum jump in a given atom. Not the speed with which quantum jumps occur, which can be random, but the frequency carried by the photons emitted when they do.

Since this frequency is given by the difference in atomic energy levels, which is the same for each atom of the same type, it is possible to use a collection of similar atoms to maintain time.

In theory, all it takes to create an atomic clock is to know the transition frequency of your particular atom. So, you can simply sit down and figure out how many swings equals one second.

In practice, you must also make sure that your atoms are in a very stable environment so that there are no involuntary changes in their energy levels.

If you can do that, you can make a clock that lacks a second only once every 50 million years!

Since the energy of each photon is very low, real-world atomic clocks need some form of amplification. Therefore, the first atomic clocks were based on the stimulated emission in the microwave range: the so-called masers.

Usually, atomic clocks use a higher power oscillator (like a quartz crystal), which is "locked" to the atomic transition frequency with an electronic feedback mechanism.

GPS itself is based on a network of satellites, each of which travels in a circular orbit approximately 20,000 kilometers above the Earth's surface. Each contains an atomic clock at the same frequency, that is, the time kept on one satellite is the time kept on another (within about one billionth of a second). Besides, each satellite continuously transmits its position and time. The classical physics of Isaac Newton defines the position.

Meanwhile, on Earth, your GPS receiver can detect the transmission signals of at least four satellites at all times, wherever you go. The receiver then defines its position by calculating its distance from these four satellites.

This can be done because the transmitted signals travel towards the receiver at the speed of light (c) and therefore cover the distance (r) between the satellite and the receiver in a time $\Delta t = r/c$. The receiver measures the discrepancy between its time and that of the satellite, then calculates the distance as $r = c \times \Delta t$. This informs the receiver that it is located on a sphere of radius r centered on the satellite's known position.

When the other three satellites are considered, the receiver determines that it is located on four spheres of known rays and centers. Two of these spheres intersect in a circle, while the third sphere intersects this circle in two points. The fourth sphere determines the position unequivocally.

All of this is based on the GPS receiver, which measures time very accurately.

So, does our smartphone need its atomic clock? Fortunately, no. The GPS receiver obtains its time from the four satellites by calculating the three positions of the coordinates (x, y, z) and the time. Solving four unknown quantities with four equations is an easy task for even the most basic computer chip.

If you want to know your precise position, say less than 1 meter, you will need something more than quantum physics. You will need to add the theory of relativity. Since the satellites themselves are orbiting at such high speeds, their atomic clocks run faster than Earth's clocks. If your receiver does not correct this, it will cause errors in the positions of a few hundred meters.

The Anti-Gravity Wheel

Suppose you have a 19 kg wheel attached at the end of a meter-long shaft. You could attempt to lift and hold that shaft with one hand and keep it horizontal. You could try to do it for fun, but I can guarantee that you won't be able to do it, unless you are a world-class bodybuilder. If however, you could manage to rotate the wheel on the shaft at a few thousand rpm (with the help of a drill, for example), you would then be able to raise it with one hand and hold it, even above your head, with minimum effort. The wheel would seem light as a feather.

How is this possible? This is due to the gyroscopic precession. Instead of pulling the wheel down to the ground as one might expect, the weight of the object creates torque that pushes it around, and therefore the wheel feels light when it turns. It can be seen that the pair vector increases the angular momentum in the same direction as the couple. If there is no angular momentum at the beginning, the new momentum oscillates in the direction of the couple. If there was an angular momentum at the beginning, the direction is changed in the direction of the angular momentum, causing the precession.

The Semiconductor

For now, we have focused on atoms (and nuclei) in isolation. Look around...

Your home is full of solid objects, from this book in your hands to the ice cubes that cool in your refrigerator. Now you will know what quantum physics has to say about solids, and you will understand how the characteristics of quantum solids can make our life better. The structure of solid matter and the atoms' arrangement in the solids derive from the number and dispositions of the electrons in each atom, i.e., from the electronic properties of the composite atoms. Although most of the solids around us are macroscopic in scale, all their physical properties are due to the electrons' movement inside them. So solid matter and all its properties are explained by quantum physics.

Solids are divided between conductors (like the copper in household wiring) and insulators (like the paper in this book). The difference is their ability to conduct electricity. If you connect a battery to a copper circuit, an electrical current will flow. Insulators, on the other hand, do not conduct electricity. There is also an intermediate category of materials that conduct electricity only sometimes: semiconductors. Quantum physics also teaches the difference between these types of solids.

Imagine two separate atoms. Take sodium, an atom with 1 electron in its external valence shell (designated 3s), and the other 10 hidden in the shells closest to the nucleus. If well separated, each atom has its own discrete but equal set of energy levels.

If you had to bring two atoms close to each other, something strange starts happening. The wave functions of all electrons begin to overlap, and the energy levels divide into two. This happens because you now have a two-atom system, and the energy levels will be a little different depending on whether the corresponding electrons have the same spin or if they have opposite spins. Your two-atom system now has a discrete set of small pairs of energy levels.

Solids have more than two atoms. When you bring so many atoms in the immediate vicinity, the individual energy levels merge into broadband due to all the overlaps of the wave function and spin-spin interactions.

Returning to the sodium atom, electron dreams from well-filled internal states (1s, 2s, and 2p) will lead to well-filled energy bands. However, what about the band corresponding to single electrons in shell 3s? The valence shell for this electron can house two electrons, even if sodium has only one, i.e., the outermost shell of an isolated sodium atom is half full.

In the solid case, this implies that even the outermost band is only half full. Furthermore, since the valence electrons of the atoms determined it, this band is called the valence band. A sodium atom can be excited in excited states, i.e., there are high energy electron shells that are occupied only when the atom has been excited.

In the solid image, this means that there are also higher stretched energy bands, and these too can be occupied if the atom gets excited. Moreover, the band closest to the valence band is called the conduction band. The occupation and proximity of these two bands are what differentiates conductors from insulators and semiconductors.

Conductors (such as sodium) have partially filled valence bands. If an electric field is applied to a conductor, the electrons can move and occupy one of the many unoccupied energy states nearby. This movement of electrons forms an electric current, better known as electricity.

On the contrary, the valence bands of insulators and semiconductors are filled. For insulators, the energy that separates the valence and the conduction bands (called bandgap) is much more extensive than typical thermal energies.

For semiconductors, the bandgap is approximately equal to typical thermal energies. Therefore, if you add some heat to the system, the electrons near the top of the valence band can be thermally excited in the conduction band, where they can flow freely. There have been no energetic states available to which electrons can be promoted to participate in electrical conduction.

Physicists have manipulated the unique properties of semiconductors to create many intelligent devices. For example, transistors – small electronic devices at the heart of any modern integrated circuit – rely mainly on the on-again, off-again nature of semiconductors. Silicon is among the most common semiconductors, hence the origin of the name "*Silicon Valley*".

The Solar Panel and the Light-Emitting Diode (LED)

We have already seen that some thermal energy (heat) is sufficient to promote an electron from the valence band of a semiconductor in the conduction band and to start the flow of electricity. We also know that electrons can undergo quantum leaps upwards by absorbing photons.

This remains true in the case of solids, just as the electrons in the conduction band produce electricity. The absorption of photons in a semiconductor is the basis of the photovoltaic cell, otherwise known as the solar cell.

The quantum properties of semiconductors can, therefore, be used to transform the free energy from the sun into necessary electricity (do you remember the photoelectric effect?).

The first solar cell was produced in 1954, at Bell Laboratories in the United States, by Gerald Pearson, Daryl Chapin, and Calvin Fuller. The cell material was not made up of a single atom type.

Physicists have discovered that the electronic properties of semiconductor materials (such as silicon) can be manipulated and enhanced by injecting a small number of impurities called a dopant. If the doping atoms have more valence electrons than the base material, you will have an "n" type semiconductor. If the doping atoms have fewer electrons, you will have a "p" type semiconductor. Solar cells and transistors are based on junctions between these different types of semiconductors.

A typical solar cell, with a diameter of about 5 centimeters and a thickness of about one millimeter, can produce about 0.2 watts of energy in full sunlight.

The matrices of 50 or more cells are electrically bonded to create panels capable of producing more useful energy amounts. Continuous improvements have been made, especially to increase the fraction of sunlight converted into electricity.

The highest efficiencies obtained today are still less than 50 %, that is, half of the sunlight still manages to heat the solar cells before getting lost in the environment.

Can solids emit light when electrons drop from higher to lower energy levels like atoms? The answer is affirmative, and this is the basis for the light-emitting diode (LED).

Today, LEDs are omnipresent in the displays of our stereos, watches, and appliances. LED lighting replaced the incandescent bulb, an innovation so precious that it won the 2014 Nobel Prize in physics.

Superconductivity

As the name already says, the only thing better than a semiconductor is a superconductor. As we have discussed in Chapter 11, superconductors take their name because these materials can carry something known as supercurrents, electric currents that flow without electrical resistance.

We have already talked about the current: it is the movement of electric charges in a conductive material.

We also talked about the resistance. It is an electrical "friction" that leads to unwanted loss of current by conversion to heat.

When connecting a battery to a simple (non-superconducting) circuit, the circuit's resistance defines the amount of current that can flow. Suppose you increase the resistance, the current decreases. The circuit's resistance draws energy from the electric current and converts it into a less useful heat form.

When you cool a superconductor, there is a low temperature below which the resistance drops immediately to zero. This is an excellent property since energy lost as heat in ordinary circuits can be preserved in their superconducting counterparts.

Quantum physics provides us with an understanding of what is going on.

Electrical resistance appears in most natural materials because every flowing electron encounter something during its journey. This could be a positive ion in the underlying material, an impurity atom, or an imperfection in the crystal structure, for example. Whenever an electron collides with one of these things, it loses some of its energy as heat.

Classical physics would say that no matter how cool you are, your collisions will remain. Therefore, it cannot explain superconductivity. Instead, it is not necessary to consider the individual properties of the particles of an electron, but the collective wavy properties of many electrons in the solid. In the superconducting state, all electrons in the material form a single coherent wave function. Once in this state, the crystal's impurities and imperfections become negligible impediments, and the current can flow without resistance.

Although zero resistance conductors appear to offer many advantages, they do have some drawbacks. The first is the need to keep the circuit fresh and pleasant, usually below about 200 ° C. It is so cold that special cryogenic liquids are needed.

Fortunately, some "high temperature" superconductors have been discovered, although we must stress that "high" is a relative term. These still need to be cooled below about -100° C to enter the superconducting state.

They are fragile materials. Therefore, they are not the best choice for electrical wiring. Therefore, it will take some time for superconducting power lines to transport electricity to our homes and reduce our electricity bills.

Superconductors have another attractive property in that they expel external magnetic fields when they are superconducting. This function is called the Meissner effect. Many technologies use this property.

For example, magnetic levitation. If a magnet is placed just above a superconducting material, the force applied by the superconductor to expel the magnetic field from its interior will allow the magnet to float in the space above. It also helped in the development of a superfast and efficient train. The "maglev" train was designed to float on some form of superconducting "guide" using magnets rather than wheels, axles, and bearings.

Today we are looking for new and improved superconducting materials, especially those that can operate at much more comfortable temperatures.

Chapter 13: Energy

Thanks to the twentieth century, we became increasingly aware that the Earth's energy sources are misused and this has led to remarkable issues related to contamination.

Some underlying concerns were focused on atomic energy, where the unavoidable radiation going with atomic procedures and the removal of radioactive waste items comprise risks, which some dreaded couldn't be controlled. This was exacerbated by few very major atomic mishaps, particularly that in Chernobyl in Ukraine, which discharged a lot of radioactive material across Europe and the past.

Recently, however, the drawn-out results of increasingly conventional strategies for energy creation have gotten clear. Chief among these is the chance of environmental change related with 'a worldwide temperature alteration': there are solid signs that the consumption of non-renewable energy sources is bringing about a progressive ascent in the Earth's temperature, which could bring about the softening of the polar ice tops, an ensuing ascent in ocean levels and the flooding of noteworthy pieces of the Earth's possessed zones. There is even the chance of a runaway procedure wherein warming would bring about additional warming until the Earth turned out to be dreadful. Notwithstanding such anticipated fiascos, there has been a fast ascent in enthusiasm for elective 'practical' types of energy creation. In this area, we will initially talk about how quantum physics assumes a job in causing the issue of dangerous global warming through the 'greenhouse impact' and how it can likewise add to a portion of the maintainable other options.

The greenhouse impact is so named because it copies the procedures that control the behavior of a glass greenhouse of the sort found in numerous nurseries. Daylight goes through the straightforward glass without being assimilated and strikes the earth and different greenhouse substances, warming them up. This procedure proceeds until the glass has gotten used to transmit as much power outwards as that of the daylight coming into it. The last procedure is helped by convection: air close to the base of the greenhouse is warmed, getting less thick and ascending to the head of the greenhouse, where it warms the glass as it cools and afterward falls back downwards.

(a) (b)

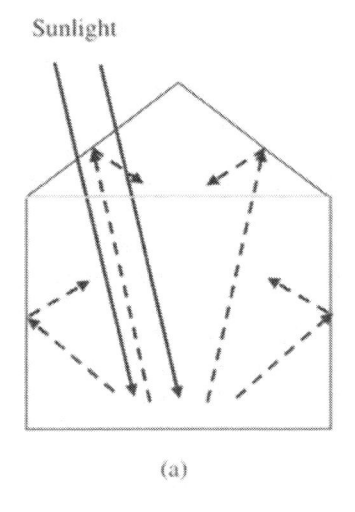

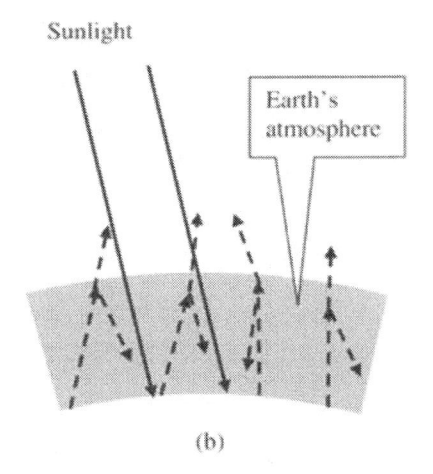

In the Figures above, Sunlight (represented by solid lines) can pass through the glass of a greenhouse (a) and warm the contents, which emit heat radiation (broken lines).

Comparable standards administer the greenhouse impact on the Earth's climate. Daylight goes through the environment to a great extent unhindered and warms the Earth's surface. The heated surface emanates warmth. A portion of this radiation is invested in the upper air and re-transmits about a portion of the reproduced energy coming back to the Earth's surface. Additionally, the excitation of such a system from its ground state can be brought about by the ingestion of a photon, however, only if its energy coordinates the distinction between the energies of the levels. Warmth radiation has a frequency of the request for 10^{-6} m (meters) and, as I clarify in somewhat more detail beneath, the energy of such a photon is like the detachment between the energy levels related with the vibration of the atoms inside the particle. Such vibrations are not promptly energized in atoms, for example, oxygen and nitrogen (the normal constituents of air), however, can be in others – specifically water and carbon dioxide. A photon that strikes one of these particles can be retained, leaving the atom in an energized state. It rapidly comes back to the ground state by producing a photon, yet this can be toward any path, and it is similarly prone to return towards the Earth all things considered to be lost to space. These 'ozone harming substances' consequently assume a comparable job to the glass in the customary greenhouse, and this procedure prompts warming of the Earth and its environment until it is sufficiently hot to re-emanate all the energy striking it. It is evaluated that without carbon dioxide, the Earth's surface's temperature would be around twenty degrees Celsius short of what it is today. In contrast, if the current measure of carbon dioxide in the air were to be multiplied, the Earth's temperature would ascend by somewhere in the range of five and ten degrees Celsius, which would jeopardize the fragile equalization on which life depends. As referenced above, water is likewise a viable ozone harming substance. Yet, the measure of water fume in the atmosphere is controlled by a harmony between the dissipation of fluid water on the Earth, prominently the outside of the seas, and its re-buildup. This is constrained by the temperature of the Earth and its air and remains, to a great extent, unaltered. However, if the Earth's temperature were to be raised substantially, the expansion in barometrical water fumes would be significant, which would prompt further dangerous global warming, then the creation of more water fumes, etc. We would have run away an Earth-wide temperature boost of a kind accepted to have happened on the planet Venus, where the surface temperature is now around 450°C.

Notwithstanding, in the present moment, at least, our anxiety isn't with water fumes, however, with different gases, for example, methane and particularly carbon dioxide. As the measure of carbon dioxide in the environment expands, the greenhouse impact prompts a corresponding rise in the Earth's temperature, an impact known as 'dangerous global warming'. Such an expansion is today brought about by human movement, especially the consumption of non-renewable energy sources. The concentration of carbon dioxide in the environment is evaluated to have expanded by around 30% since the mechanical movement started in around 1700, and is as of now expanding by about 0.5% every year, which, if it proceeded, would prompt a multiplying in around 150 years and an ensuing an Earth-wide temperature boost of somewhere in the range of five and ten degrees Celsius.

In the figures in the next page, the electronic charge clouds in the carbon, and (a) illustrates the dioxide molecule. In a molecule, the atoms can move as if they were connected by springs, as shown in (b). In the molecule of carbon dioxide, the carbon at the center carries a net positive electric charge, and the two outer oxygen atoms are negatively charged. When the molecule gets an electric field applied to it, oppositely directed forces are applied to the oxygen and carbon atoms, which in turn respond as illustrated in (c), so exciting the vibration shown in (b).

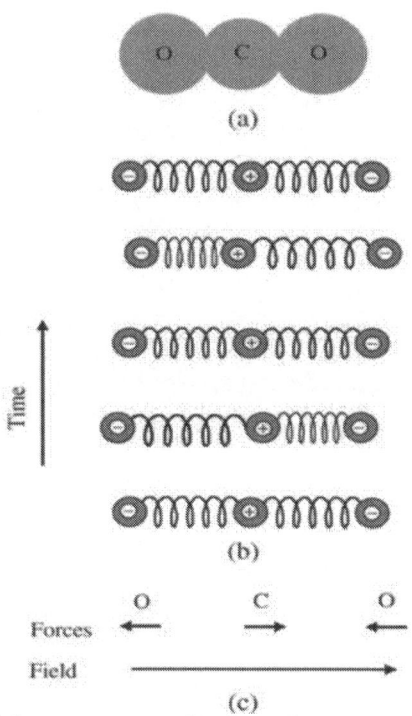

We can understand in somewhat more detail how quantum physics guarantees that gases, for example, carbon dioxide go about as ozone harming substances while the more typical constituents of air – nitrogen, and oxygen – don't. In any case, the energy of a photon related to the warmth transmitted from the Earth's surface is around multiple times not as much as this, so an alternate sort of procedure must be related to the assimilation of this low-energy radiation. The key point here is that in a particle, the nuclear cores can be made to vibrate comparative with one another. This implies if we had the option to move the cores somewhat away from this harmony position, the energy would be raised so that if we currently discharged them, the cores would move back towards the balance point, changing over the abundance energy into dynamic energy related with their movement. They would then overshoot the harmony point, slow down and return, and this vibrational movement would proceed inconclusively except if the energy were lost here and there. In this sense, an atom carries on as though the cores were point masses associated with springs, experiencing swaying as the springs stretch and agreement. The figures above show this for the instance of the carbon dioxide particle, which comprises a carbon molecule bound to two oxygen atoms in a direct design. Warmth emanated from the Earth's surface has a scope of frequencies, which envelop the vibration frequencies of the gases in the atmosphere, including those of ozone-depleting substances, such as carbon dioxide.

The above applies the same amount of nitrogen and oxygen as it does to carbon dioxide and water. Hence, despite everything, we need to understand why heat radiation can actuate vibration in the last two gases yet not the previous two. Since the two hydrogen atoms are indistinguishable, this particle is symmetric, and the two separating charge veils of mist are additionally indistinguishable. Notwithstanding, this isn't accurate on account of more complex atoms. Considering the most minimal energy condition of carbon dioxide specifically, things being what they are, the absolute charge in the cloud encompassing the focal carbon atom is somewhat less than six electronic charges thus doesn't completely adjust the charge on the carbon core, while the charge encompassing every oxygen molecule is somewhat more than that corresponding to the total of eight related with a free oxygen particle. The net impact of this is, even though the total electronic charge on the particle adjusts the absolute atomic charge, every oxygen molecule conveys a little net negative charge, and an adjusting positive charge is related to the carbon atom. We currently consider what happens when the particle is exposed to an electric field coordinated along its length. This permits the ingestion of energy, which is then re-transmitted in an arbitrary way, thus prompting a greenhouse impact. A similar procedure happens when an applied field is opposite to the particle line. The carbon atom moves a single way and the two oxygens in the other, which makes the atom twist for this situation. This likewise prompts a greenhouse impact for radiation of the suitable recurrence.

Why then does a comparable impact not emerge in particle-like oxygen or nitrogen? The explanation is that such a particle contains two indistinguishable atoms, which should consequently either be impartial or convey a similar net charge. In either case, they can't be pushed in inverse ways by an electric field, so a vibration can't be set up by an electromagnetic wave, and such a gas can't add to the greenhouse impact.

If quantum physics assumes a role in creating the greenhouse impact and its related issues, can it likewise help us maintain a strategic distance from and resolve them? We realize that atomic responses are represented by quantum physics laws and produce no carbon dioxide or other ozone-depleting substances. In this manner, the age of atomic energy (both parting and combination) does not contribute to the greenhouse impact. We have seen that atomic energy may have issues of its own, and it surely has had an awful press since around

1980. However, a few naturalists have been changing their suppositions as of late. Other 'green' types of intensity incorporate wind power, wave power, and solar power. Sunlight based energy comes in two principal structures. It may very well be used to warm local high temp water systems (for instance), and again, there is nothing specifically quantum about this procedure. However, it can likewise be used to deliver power in 'photovoltaic cells', whose exhibition relies upon quantum impacts.

Chapter 14: Philosophical Implications

In order to reveal the logical system of quantum physics, it is important to refer to the empirical status of the relationships between entities in the physical world. This can be exposed through different experiments that question whether a physical reality has a reflection in the quantum state or not and vice versa.

Before the quantum era, science lived on decisive pronouncements on causes and effects of motions: well-defined objects moved along precise trajectories, in response to the action of various forces. But the science that we now call classical, emerged from the mists of a long history and duration until the end of the nineteenth century, overlooked the fact that each object was made up of a gigantic number of atoms. In a grain of sand, for example, there are several billion of them.

Before the quantum era, anyone who observed a phenomenon was like an alien from space, who looked at the Earth from above and noticed only the movements of large crowds of thousands and thousands of people. Maybe they saw them marching in compact ranks, or applauding, or hurrying to work, or scattering in the streets. But nothing they observed could ever prepare them for what they would see by focusing their attention on individuals. On an individual level, humans showed behavior that could not be deduced from that of crowds - things like laughter, affection, compassion, and creativity. Aliens, perhaps robotic probes or evolved insects may not have had the right words to describe what they saw when they observed us closely. On the other hand, even today, with all the literature and poetry accumulated over the millennia, sometimes we cannot fully understand other human beings' individual experiences.

At the beginning of the 20th century, something similar happened. The complex building of physics, with its exact predictions about the behavior of objects, i.e. crowds of atoms, suddenly collapsed. Thanks to new, sophisticated experiments, conducted with great skill, it was possible to study the properties not only of individual atoms but also of the smaller particles of which they were made. It was like going from listening to an orchestral ensemble to quartets, trios, and solo pieces. And the atoms seemed to behave disconcertingly in the eyes of the greatest physicists of the time, who were awakening from the sleep of the classical age. They were explorers of an unprecedented world, the equivalent of the poetic, artistic, and musical avant-garde of the time. Among them were the most famous: Heinrich Hertz,

Ernest Rutherford, J. J. Thomson, Niels Bohr, Marie Curie, Werner Heisenberg, Erwin Schrödinger, Paul Dirac, Louis-Victor de Broglie, Albert Einstein, Max Born, Max Planck, and Wolfgang Pauli. The shock they felt after poking around inside the atoms was equal to what the crew of the Enterprise must have experienced when they first encountered an alien civilization found in the vastness of the cosmos. The confusion produced by the examination of the new data slowly stimulated the first, desperate attempts by physicists to restore some order and logic in their science. At the end of the twenties of the last century, the atom's fundamental structure could be said by now known in broad lines, and it could be applied to the chemistry and physics of ordinary matter. Mankind had begun to understand what was happening in the new, bizarre quantum world.

But while the crew of the Enterprise could always be teleported away from the most hostile worlds, the physicists of the early twentieth century did not go back: they realized that the strange laws they were discovering were fundamental and were the basis of the behavior of all matter in the universe. Since everything, including humans, is made of atoms, it is impossible to escape the consequences of what happens at the atomic level. We have discovered an alien world, and that world is within us!

The shocking consequences of their discoveries upset not a few scientists of the time. A bit like revolutionary ideologies, quantum physics consumed many of its prophets. In this case, the ruin did not come from political machinations or conspiracies of adversaries, but from disconcerting and deep philosophical problems that had to do with the idea of reality. When, towards the end of the 1920s, it became clear to everyone that a real revolution had occurred in physics, many of those who had given it the initial impetus, including a figure of the caliber of Albert Einstein, repented and turned their backs on the theory they had contributed significantly to creating. Yet today, well underway in the 21st century, we use quantum physics and apply it to a thousand situations. Thanks to her, we have invented transistors, lasers, atomic energy, and countless other things. Some physicists, even prominent ones, continue to use all their strength to find a version of quantum mechanics softer for our common sense, less destructive than the common idea of reality. But it would be good to reckon with science, not with some palliative.

Before the quantum era, physics had managed very well to describe the phenomena that happen before our eyes, solving problems in a world made of stairs firmly resting on the walls, arrows and cannonballs launched according to precise trajectories, planets orbiting and rotating on themselves, comets returning to the expected time, steam engines doing their useful work, telegraphs and electric motors. In short, at the beginning of the twentieth century, almost every observable and measurable macroscopic phenomenon had found a coherent explanation within the so-called classical physics. But the attempt to apply the same laws to the strange microscopic world of atoms proved incredibly difficult, with deep philosophical implications. The theory that seemed to come out, the quantum theory, went completely against common sense.

Our intuition is based on previous experiences, so we can say that even classical science, in this sense, was sometimes counterintuitive, at least for the people of the time. When Galileo discovered the laws of ideal motion in the absence of friction, his ideas were considered extremely daring (in a world where no one or almost no one had thought to neglect the effects of friction). But the classical physics that emerged from his intuitions managed to redefine common sense for three centuries, until the 20th century. It seemed a solid theory, resistant to radical changes, until quantum physics burst onto the scene, leading to an existential shock like never before.

To understand the behavior of atoms, to create a theory that would agree with the seemingly contradictory data that came out of the laboratories in the thirty years between 1900 and 1930, it was necessary to act radically, with a new audacity. The equations, which until then calculated with precision the dynamics of events, became tools to obtain fans of possibilities, each of which could happen with a given probability. Newton's laws, with their certainties (so we speak of "classical determinism"), were replaced by Schrödinger's equations and Heisenberg's disconcerting mathematical constructions, which spoke the language of indeterminacy and nuance.

In the 21st-century quantum mechanics has become the backbone of all the research in the atomic and subatomic world, as well as wide sectors of material sciences and cosmology. The fruits of the new physics make hundreds of billions of dollars every year, thanks to the electronics industry. Many follow the improvements in efficiency and productivity made possible by the systematic use of quantum laws. However, some physicists, a bit rebellious, driven by the cheers of a certain type of philosophers, continue to seek a deeper meaning, a principle hidden within quantum mechanics in which determinism is found. But it is a minority.

As we approach the new atomic territories, everything that intuition suggests becomes suspicious and the information accumulated so far may no longer be useful to us. Everyday life takes place within a very limited range of experiences. We do not know, for example, what it feels like to travel a million times faster than a bullet, or to endure temperatures of billions of degrees; nor have we ever danced in the full moon with an atom or a nucleus. However, science has compensated for our limited direct experience of nature and made us aware of how big and full of different things the world out there is. To use a famous metaphor, we are like chicken embryos that feed on what they find in the egg until the food is finished, and it seems that our world must end too; but at that point, we try to give the shell a shot, we go out and discover an immensely larger and more interesting universe. Among the various intuitions typical of an adult human being there is the one that the objects that surround us, whether they are chairs, lamps, or cats, exist independently from us and have certain objective properties. Based on what we study in school, we also believe that if we repeat an experiment at various times (for example, if we let two different cars run along a ramp), we should always get the same results. It is also obvious, intuitive, that a tennis ball passing from one half of the court to the other has a defined position and speed at all times. It is enough to film the event, i.e. obtain a collection of snapshots, to know the situation at various moments, and to reconstruct the overall trajectory of the ball.

As we have seen and will keep seeing, to understand quantum theory we must enter a new world. It is undoubtedly the most important fruit of the twentieth century's scientific explorations, and it will be essential for the whole new century. It is not right to let only professionals enjoy it.

Even today, at the beginning of the second decade of the 21st century, some illustrious scientists continue to search with great effort for a more "friendly" version of quantum mechanics that less disturbs our common sense. But these efforts so far do not seem to lead to anything. Other scientists simply learn the rules of the quantum world and make progress, even important ones, for example adapting them to new principles of symmetry, using them to hypothesize a world where strings and membranes replace elementary particles or imagining what happens at scales billions of times smaller than those we have reached so far with our instruments. This last line of research seems the most promising and could give us an idea of what could unify the various forces and the very structure of space and time.

It is important to highlight that there is no magic in Quantum Physics. It complies with the laws of the universe and doesn't suspend the rules of common sense. The essential rules that direct physics are unblemished: it preserves energy and doesn't contradict our logic, there is still an increase in entropy, and nothing is faster than the speed of light.

In Quantum Physics, there are endless attributes that resist our systematic instinct such as probabilistic counts, the indeterminacy of states, non-local effects. However, this is not conflicted with good judgment or thinking. Therefore, Quantum Physics cannot be used for divination or perceptiveness or any of the numerous parts of magic and mysticism.

I aim to make you appreciate the disturbing strangeness of quantum theory, but above all the profound consequences it has on our understanding of the world. I believe that the uneasiness is mainly due to our prejudices. Nature speaks in a different language, which we must learn - just as it would be good to read Camus in the original French and not in a translation full of American slang. If a few steps give us a hard time, let's take a nice vacation in Provence and breathe the air of France, rather than stay in our house in the suburbs and try to adapt the language we use every day to that very different world. In the next chapters we will try to transport you to a place that is part of our universe and at the same time goes beyond imagination, and in the next chapters, we will also teach you the language to understand the new world.

Chapter 15: Law of Attraction

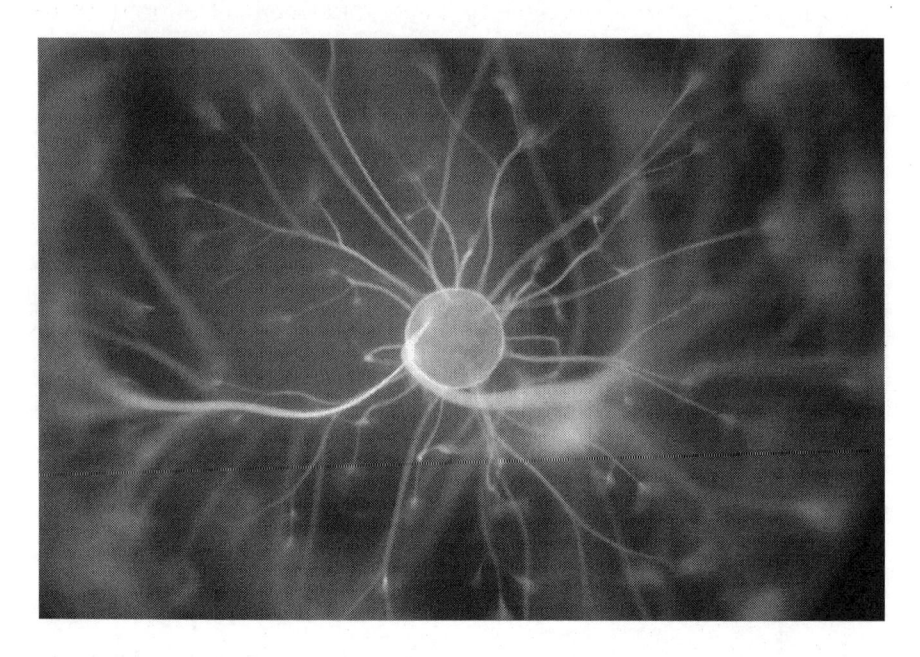

If this was a motivational book, you would be reading something saying that the Law of Attraction is a powerful universal law that affects every facet of your life. It is the secret to being happy, living a successful life, and being able to fulfill what you want. The tune of your frequencies and vibrations is reflected in the universe.

The quality of your life, your wealth and abundance (or lack thereof), and your degree of success, determine your understanding of the law of attraction. It also defines whether it is directly applied or not.

When the Law of Attraction is given a voice, it says: "You attract into your life whatever you give your attention, focus, and energy to." This is true whether what you attract is desirable or undesirable. Think of yourself as a large magnet. On that magnet, there is a control dial, which has settings ranging from disastrous to wonderful. The dial set is reproduced in your life. It determines and defines the opportunities, the people, the events, the circumstances, and the coincidences that show up in your life.

As stated in quantum physics, everything in this universe is energy: starting from yourself to the chair you are sitting on, to the clothes you are wearing, and even your thoughts and feelings.

The prerogative "everything is energy" renders our experience of life paradoxical. If we think that we are solid and that the chair we are sitting on is also solid, we need to change this and hence keep in mind that everything is energy. Energy vibrates, and different qualities of energy vibrate at different rates. While energies of the same frequency attract each other and resonate, those of different frequencies repel each other.

Your thoughts and feelings are energy. They are absurd and contradictory as they penetrate both time and space and attract similar energies and repel dis-similar energies. These energies are incarnated later in people, events, circumstances, and coincidences.

In addition, you are the one who controls your reality when you believe that the Law of Attraction responds to the thoughts that you have at all times. Every experience in your life is driven by the Law of Attraction and reflects your inner thoughts. It doesn't matter if you recall something from the past, not living your present, or projecting something into your future. The thoughts you are focused on in the powerful present triggers a vibration that exists within you. Then, the Law of Attraction responds to it at the present.

When you come across something you would like to experience, and you strongly think "I would like to have that" the attention you give to it will turn it into an experience. On the other hand, when you focus on something you do not want to experience and you strongly shout, "No I don't want this!" your attention will invite this to you and turn it into an experience. The universe is based on attraction and as a result, there is no exclusion or exception. Your attention to the universe includes it into your vibration and when you hold a thought in your awareness or your attention too long, the Law of Attraction will make it happen and turn it into experience because no is not an answer.

Plainly, negative thoughts and experiences are immediate calls to experience something. As an example, when you focus on something and shout "Go away, I don't wish to experience this", you are attracting the event since no does not exist in the Universe of attraction. In other words, your attention is saying, "Yes, I want to attract this event, even if this is something I do not want." The good news is that in our physical reality of time and space, such occurrences do not manifest into experiences quickly. The amazing buffer of time exists and is situated between when you start thinking of something and the actual time it manifests. The power that offers you the opportunity to focus your attention more on the way things wish to manifest is the buffer of time. Long before things manifest as you first start thinking of them, you will be able to know by the way you feel if this is something you wish to manifest or not. If you keep focusing on it, no matter if this is something you want or not, it will happen in reality.

The magnetic power of the Law of Attraction influences the Universe and works on similar thoughts in terms of vibration to carry them to you. It is a correlative relationship between your activated thoughts, focused subjects on the one hand and on the other persons, circumstances and experiences that are in constant response to the Law of Attraction. Although it functions transcendently, it can be exemplified as a tunnel that includes a powerful magnetism to hold your thoughts' vibrations.

Based on your feelings and emotions, you can be aware if you are shaping in your mind the things that you want or un-want. In fact, the Law of Attraction will always respond and act on the vibrations you are sending. At first, when you hear about the Law of Attraction, and you assimilate that you can attract things by controlling your thoughts, you start to monitor each thought to presumably avoid negative things. Yet, it is noteworthy that it is difficult to monitor all your thoughts because people have so many things to think about and the Law of Attraction will bring even more thoughts to us.

Quantum Physics and the Law of Attraction

Apparently, it seems difficult to understand how the universe works. How you can get what you want, and how sometimes, you just don't seem to get it. The Law of Attraction and Quantum Physics work together to create equilibrium in the universe. It is important to understand both of them so that you can understand how the Universe works.

To begin with, the Law of Attraction – along with Quantum Physics – boils down to a very basic aspect that you need to understand to make good use of the Law of Attraction. It is important to remember this fact as you deal with the Law of Attraction so that you know that you can make the most of the law and what it means.

It is ironically needless to say that 'Like Attracts,' you look at exactly how it sounds. Thus, if you focus on the way you are, your attitude, hopes, and dreams will attract similar things to you. It all depends on the energy you diffuse to the universe.

Sometimes, you are overwhelmed by the feelings of anger, you are upset or running late, the more upset and frustrated you were about the day, the later you seemed to be running. The more you dwell on being late, irritated, and angry, the more you see that you give yourself more reason to be upset, frustrated, and late. Then think about a good day you've had in your life—a day when everything seems to be going your way. You might be excited and happy, and there seems to be nothing that can bring you down. The more you concentrate on these happy and excited emotions, the more you're going to be happy and excited.

This is the fundamental idea behind the Law of Attraction - Like Attracts Like. It can be summarized as: the more you concentrate on good things and positive things, the more the World gives you good things and positive things.

This concept has been around for a long time, but it has only recently become popular, as more and more people begin to understand that the Law of Attraction is Quantum Mechanics, a theory of how the universe functions. Quantum Physics teaches that nothing is set, that there are no limits, that everything is vibrating energy. This Energy is under the control of our feelings. It is shaped, formable, and moldable. It's different than simply wishing and hoping, it boils down to believing. To make the Law of Attraction work for you, you must believe that the Universe will send you the things you want.

The law of attraction is one of the simplest laws when fully grasped. It is advantageous because it helps you manipulate everything you've ever dreamed of.

The Law of Attraction is something that tells a person to draw things to themselves by concentrating on certain things. It has a relationship with Quantum Mechanics, which explains that there is nothing definite and there are no limits. According to Quantum Physics, all is made up of vibrating energy. The Law of Attraction and Quantum Physics are therefore both related and, in fact, interrelated.

Unlike Newton's classical physics which states that the universe is made of solid building block, Quantum Physics and the Law of Attraction assume that the universe is fluid and constantly changing to respond to people.

The Quantum Law of Attraction, therefore, is that because everything is always evolving and fluid - and, in reality, because the Universe is made up of these dynamic and changing energies - everything can be attracted to any person, simply by concentrating or focusing on it. According to Quantum Physics, every person is part of the creation of the universe. Each person focuses on issues and attracts them, and according to the issues that they are concentrated on, those items are brought to each person. Therefore, the world is affected by our feelings. In reality, it's not something that's set in stone. It's movable and influenced by people's thoughts and what they believe in.

In reality, bringing things to an individual is the only way to obey both Quantum Physics and the Law of Attraction at the same time.

Focusing on the things you want and keeping them at the forefront of your mind is the best way to make sure you're motivated to do those things.

If you want to fulfill your dreams and get out of the feeling of being trapped, you need to believe that everything in this universe is energy and that this energy resides in a state of possibility. You have to allow the rule of attraction to be enforced to achieve success. Remember, we are the builders of the universe. According to Newton's classical physics, the universe is made up of discrete building blocks. These blocks are solid, and they cannot be changed.

Quantum physics explains that there are no separate parts of the universe. All exists in the form of fluid and tends to change from time to time. Physics imagine this world as a deep ocean of energy that keeps coming into existence and disappearing out of this universe.

Chapter 16: Quantum Physics and Health

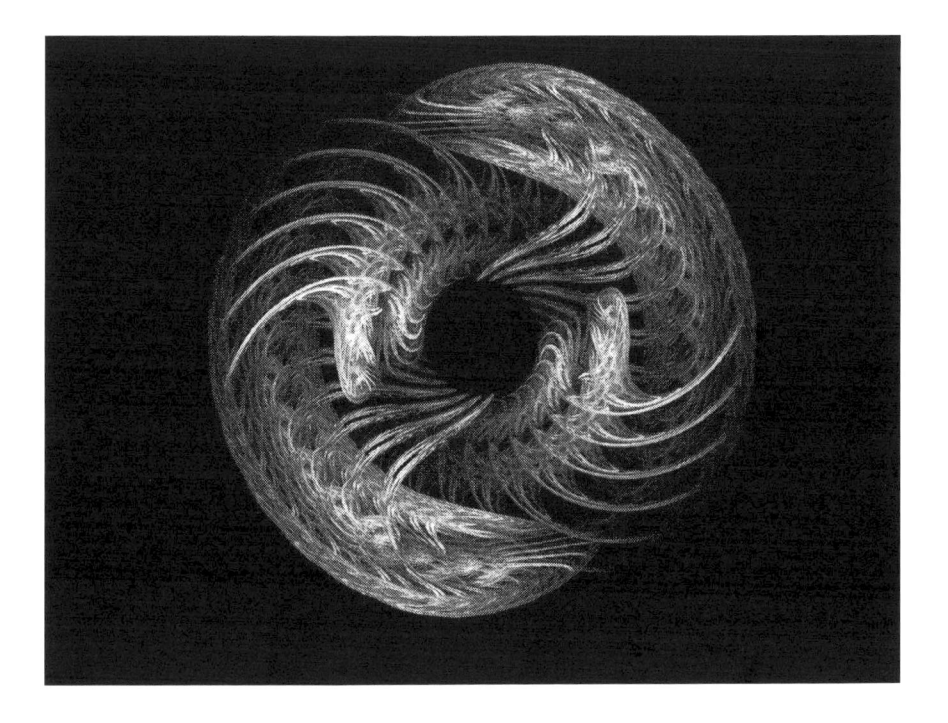

Quantum physics has had and still has a significant role in contributing to the development of tools and techniques aimed at improving the health of human beings. In the next pages, we will look at some of its applications.

Improved disease screening and treatment

Utilizing a new technique recognized as the bio-barcode test, experts can now use gold nanoparticles to detect disease-specific signs, or "biomarkers," in our blood, which are visible using MRI imaging and have special quantum properties that allow them to bind to disease-fighting cells. Such nanoparticles of gold are completely safe for human consumption. This approach is also less costly, more versatile, and more accurate than traditional alternatives.

Mikhail Lukin, a Harvard physics professor and pioneer in quantum optics and atomic physics, also focuses on engineering nanoscale diamond particles for similar purposes. He aims to potentially use non-toxic diamond particles to take images of human cells from within and to diagnose illness without exposing patients to radiation.

By allowing ultra-precise measurements, quantum sensors can also improve the MRI machine itself. Instead of the whole body, a new form of quantum-based MRI could be used to look at one molecule or groups of molecules, offering physicians a far more accurate image.

Many quantum-based approaches for treating diseases are also being developed. For instance, a gold nanoparticle can be "set" to build up in tumor cells only, allowing accurate imaging as well as tumor laser destruction without destroying healthy cells.

No more needles

University of York researchers have developed a pad that can be added to the skin to administer tailored treatments without hypodermic needles. The pad, dubbed the Nanject, will be used without damaging healthy cells to distribute cancer drugs.

Here's how it goes: before being inserted into the bloodstream, the nanoparticles have antigens where they bind to cancer cells. The patient is then placed in an MRI machine, which causes the particles to heat up and destroy the cancer cells. The ions cool back down when the unit is switched off and can be withdrawn from the body without hurting the user.

Needle-phobic patients can also be excited about this kind of progression: the Nanject patch substitutes one syringe with a lot of tiny ones consisting of polymers that distribute the drug via the hair follicles. However, the nanotech drug delivery route has another, perhaps more important benefit: it removes some of the toughest barriers to medication distribution, especially in remote and impoverished areas. There is no need for a qualified nurse or surgeon to deliver drugs with a patch; it can be done by anyone through a procedure that is as easy as putting on a band-aid.

Nanotech drug delivery also authorizes lower doses, as nanoparticles are not consumed like pill-based medicines by stomach acid. Finally, medications such as the Nanject can help prevent disease transmission by unsterilized needles—a major problem in developing nations.

Hacking human biology

Quantum mechanics can provide us with more knowledge about human biology beyond better disease detection and highly targeted, needle-free therapies.

Australian scientists have recently discovered a way to investigate a living cell's inner workings using a new method of laser microscopy based on the concepts of quantum mechanics. And we can use quantum computers to sequence DNA quickly then solve other healthcare challenges with Big Data. This opens the possibility of specialized treatment based on the unique genetic structure of people.

More secure health data

For obvious reasons, people want to protect their health data, so it's important to consider all the ways it can be compromised. For example, in the future, hackers may become able to intercept messages retroactively.

A growing number of cyber security companies use quantum phenomena' peculiar properties to secure data in an ultra-safe way.

Using quantum entanglement is one of the most practical applications of the technology to date. Quantum cryptography prohibits anybody other than the intended recipient from accessing the results. Some companies already provide banks and governments with protection; over the next few years we can expect to see such protection to expand to the healthcare sector, too.

Innovations based on quantum mechanics principles have the potential to affect health care at almost any level, from diagnosis and treatment to data storage and transmission. We need to keep a close eye on quantum science and healthcare, a field that will benefit from increased support for R&D. We are on the cusp of some fascinating developments, and we should all be educating ourselves on how quantum science in the not-so-distant future can change healthcare and ultimately drastically improve our lives.

Quantum mind, meaning, and medicine mind as slayer

Quantum mind (sometimes referred to as quantum consciousness) is the belief that consciousness involves quantum processes, as opposed to the concept of conventional neurobiology in which the operation of the brain is entirely classical, and quantum processes play no role in computation.

While many who strive at a proposition of quantum consciousness are false and mistaken by naively believing that the weirdness of quantum mechanics is similar to the strangeness of consciousness, more advanced theories of quantum consciousness are an attempt to solve the "combination question," that is the problem that describes how a network of classical neurons can combine to form a single subject of experience.

Nevertheless, no experimental evidence of computationally important quantum processes is currently available in the human brain, partly due to the technical difficulty of studying the brain at an adequate spatial and temporal granularity.

Quanta and consciousness

Not every quantum mechanical interpretation presumes that quantum collapse occurs, but one competing theory about how it occurs is that consciousness causes collapse if it does. Since these ideas focused on consciousness were developed, this form of quantum physics has been drawn by people who want to believe that consciousness is in some way special.

Several followers to quantum physics focused on supernatural consciousness were woo-misters and pseudoscientists who often suggested costly solutions to problems. This also attracted respected scientists such as Eugene Wigner to quantum consciousness, a view which Wigner repudiated. It is difficult for laypeople to see at the present level how far quantum consciousness is a reasonable theory, and how much it is wishful thinking.

However, it should be lauded that barely substance dualism is true, which the majority of the scientists doubt. It should be possible for conscious minds to faint wave functions just as much as unconscious photodetectors can. If this were not the case, it would mean our minds are composed of some non-physical material that does not make up implicit photodetectors.

Chapter 17: String Theory and the Theory of Everything

All throughout this book, you've heard of the term "string theory," but you might not yet have a precise idea of what it means. As it turns out, this is the case for many people, and not by their fault, either — string theory is just really, really confusing.

String theory is so named because it was created to describe the nature of quantum objects as something like a vibrating string, and those objects were measured based on those vibrations. Without getting too deeply into it — string theory involves a lot of work with subatomic particles, which are the smaller particles that makeup atoms (like quarks, for example) — string theory was essentially formulated to help explain strange interactions that sometimes happened between some of the known subatomic particles. These particles would sometimes act like they were bound together by strings. Thus, string theory was formed.

Now, one of the most interesting parts of this early string theory was that these weird, string-based interactions worked out some mysterious math. For one, the vibrations in the strings predicted the existence of a certain particle: the graviton. It's the only quantum theory that has been able to successfully do this so far. Theoretically, a graviton is a particle that causes gravity, and one of string theory's biggest advantages here is that its graviton is shaped like a donut. That may sound strange, but that same shape prevents many of the mathematical anomalies that arise when trying to envision gravity as a particle. However, with today's science, we still can't prove that the graviton even exists. Gravity still presents many mysteries to us.

The string theory was able to successfully describe gravity as its quantum particle this way. However, string theory has many limitations that keep it from being an attractive (or logical) solution to the quantum universe's mysteries. For one, the final proposal of string theory requires ten dimensions to function properly, but we've so far only observed four dimensions in our existence. To remedy that, physicists and scientists alike have tried to make string theory work in its prerequisite ten dimensions, then remove the extra dimensions that don't apply to our universe. However, none have yet been successful. Besides, there are several other versions of string theory – there isn't just one. Bosonic string theory was one example, and it required twenty-six dimensions!

As we will see more in details in the next chapter, Ed Witten managed to combine many of the varied string theories into M-theory. However, M-theory required eleven dimensions, so while it managed to unify many of the past string theories into one neat package, it still didn't seem anywhere near reasonable science.

In addition, the "theory of everything," as its name implies, is an all-encompassing theory. It explains in a way or another how quantum physics and quantum particles worked. Although it's one of today's physicists and scientists' premier goals, we haven't formulated a working theory of everything yet. However, string theory makes a compelling case for becoming or at least supplementing, a potential theory of everything.

It seems that string theory makes a very neat argument to explain the nature of particles which were covered in chapter 7 (Waves and Particles). Envision a guitar string in slow motion: the string moves in a wave shape, just like light, or sound. The tension (how taut the string is pulled) of the string also controls what note any string on a guitar plays. If you pluck the string, it will vibrate. This is the prevailing theory behind the strings: each string has a different frequency (like the note of a guitar string) that it vibrates at, and each string also has a different length, which changes the number of notes that the string could play. Think about how, when playing the guitar, moving your fingers down the neck, shortening the string, will play higher notes while moving your hand up the neck will lengthen the string and play lower notes. Interestingly enough, the intervals between these possible "notes" are also defined by a familiar variable: Planck's constant. Unfortunately, as convenient as this might seem, there's one question that scientists seem to be unwilling (or unable) to answer about string theory. What are the strings made of? Answers that have been tried are "pure mass" (whatever that means), irregularities in the fabric of reality, and the "energy" of "existence." The best answer, however, is, "The answer is irrelevant because they're there."

One of the biggest downfalls of string theory is that, because it's so complicated and exists in so many additional dimensions (that we can't even find), we can't test it. However, we also can't rule it out as entirely wrong, either. This creates a bit of a catch twenty-two: should scientists and physicists keep pouring man-hours and research into a theory that we technically can't even test, or should we just give up on it, and risk sidelining a potential theory of the universe? There's no good answer to this question, but it's one of the pitfalls that plague string theory and those interested in it.

Despite its shortcomings, though, many scientists still believe that string theory is the way things are – or, at least, it's a step to figuring it out. However, until we find six or seven more dimensions, we'll have to wait on that.

Chapter 18: M-theory

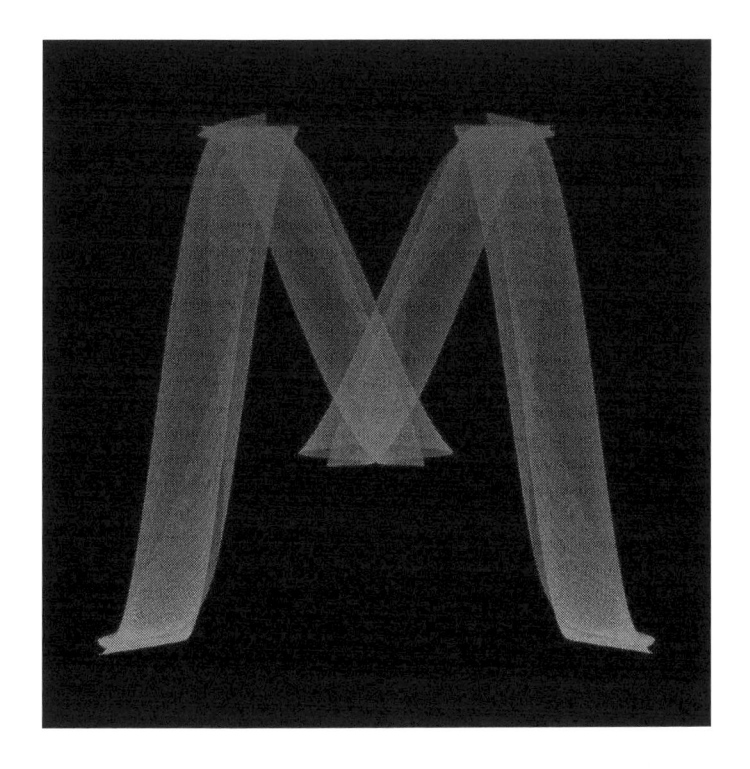

Whenever there is a string theory breakthrough, it generates excitement and shock waves throughout the theoretical physics community. Scientists and researchers generate plenty of papers and literature about the breakthrough. M-theory is one such breakthrough because it appears to explain the origin of the strings themselves. How is this possible?

With M-theory, a long series of puzzles regarding string theory is solved. Many of these puzzles have dogged string theory since the beginning. Yet M-theory is so encompassing that it may even force a change in the name of string theory. How can one theory make such a dramatic change? Simply because the M-theory is not based purely on strings. While theorists hesitate to definitively state that the theory has been proven correct, the M-theory marks a significant breakthrough that has already begun to reshape this particular field of study. One of the prominent theorists in M-theory is Edward Witten. He conjectured that M-theory is grouping String Theory and the Theory of Everything.

Besides, when it comes to an understanding of how all the forces of the universe come together, scientists continue to look for the unified field theory. Scientists can believe that such a theory exists, but the question is whether these individuals can think far enough outside of the box to finally explain this theory.

Einstein himself believed that nature has shown us pieces of this particular puzzle, but that it is up to our researchers and scientists to put all these pieces together. He spent the last 30 years of his life looking for the equation that would be the theory of everything, allowing all the forces of the universe to be united in a single equation. Other expeditions and dreams about unified field theory have failed to adequately convey all the forces into one solid theory. Yet, for every fail, another trail presents itself for scientists to follow. The superstring theory is preeminent, and at the moment, the only candidate to be considered for the theory of everything. While surviving all the mathematical challenges thrown at it, primarily because of the radical nature of the theory, which is based on extremely tiny strings that are vibrating in 10-dimensional space-time, this theory has swallowed up even Einstein's theory of gravity because string theory requires gravity, which is new to these theories.

Still, with every quantum theory, no matter how radical, there seems to be a weak spot that becomes obvious to the researchers and scientists over time. The weak spot for the string theory is that they can't seem to probe all the model's solutions, particularly when it comes to examining the non-perturbative region. But this is, for the most part, a critical area since our universe and all of its fantastic parts, may lie in this non-perturbative region.

Thus, M-theory is a breakthrough that uses a powerful tool called duality to solve many of the issues with superstring theory. As a result, we have a potential answer to the question of where the strings come from. For string theorists, it is somewhat embarrassing to have five self-consistent strings. Still, all of them do essentially the same thing, unite two fundamental physics theories, quantum theory and, of course, the theory of gravity. Yet each of these theories appears to be unique because they are each based upon diverse symmetries with unusual names.

String theories are not the only theories that contain supersymmetry. This type of mathematical symmetry is also part of a few other areas, such as changing light into electrons, then into gravity. This is also a symmetry that exchanges particles that have a half-integral spin with those that have an integral spin.

It has been determined that 11 different dimensions have alternative super theories, which are based upon the idea of membranes and point particles. Lower dimensions are crowded with super theories using membranes in diverse dimensions. There are also p-branes, part of the p-dimensional case, but these have been determined to be very hard to work with, and researchers have often abandoned their work with them as a virtual dead end. With all these possibilities, how can supersymmetry make them all work?

Scientists now realize that these are all just different facets of the exact theory. Thus M-theory carries the distinction of uniting all these string theories while including the p-branes. In fact, these are all limits on the same theory. Imagine for a moment that you aren't able to see. As you feel around, you come across the tail of an animal. This might appear to be like the string or a one-brane. Then you feel an ear or the membrane. Finally, you feel a leg or a three-brane. But are any of these parts stand-alone pieces? No, they are all part of the same thing. Thus, researchers have found that perhaps it is the researchers' viewpoint that need to be adjusted to view the strings in a different light.

These lead to other concepts, such as perturbation theory, non-perturbation solutions, and even duality. So, let's look at each of these concepts to gain a deeper understanding of how they work by looking at how they each became part of the world of physics.

Duality: The Challenge of Equivalent Theories

When it comes to the M-theory, understanding duality is critical. This duality transpires when two distinct theories can demonstrate their equivalence in various interchanges. An example is when James Clerk Maxwell, Cambridge University, reversed the roles of electricity and magnetism within his equations.

His equations are standards that direct light, x-rays, motors, radars, and even the televisions we have all watched at one time or another. What makes these equations even more unique is that they remain the same, even with switches between the magnetic and electric fields, along with switching the two different charges. As a theory lacks the ability to be solved precisely, scientists use an approximation scheme to solve them. Adding these different contributions, the system being studied gets perturbed, thus demonstrating the perturbation theory.

If you have ever shot an arrow, then you have demonstrated the perturbation theory just by aiming your arrow. Why is this the case? Because every motion that you make with your arms allows you to gradually line up with the bull's eye.

Non-perturbation, on the other hand, adds larger motions or contributions, and these eventually make the non-perturbation region meaningless. But thanks to duality, the minor side of the equation (perturbation) is simple to solve and is identical to the superior side (non-perturbation). So we can solve both by solving one side.

For those who study string theories, this idea of duality led them to apply it to the original 5 string theories. The results were very interesting, to say the least, as the 5 string theories were reduced to 3, thus showing that these researchers might be on the right track.

Alphabet Duality: Understanding S, T, and U

So, who were the first researchers to explore duality in terms of string theory? K. Kikkawa and M. Yamasaki of Osaka University were the first to step into this arena in 1984. These two researchers demonstrated that if one of the extra dimensions was curled into a circle with an R radius, then it would be the same as if you curled up the same dimension, but with a 1/R radius instead. Called T-duality or

R<->1/R, it could be applied to superstring theories. When that happened, the 5 superstrings were brought down to just 3. How is this possible? Simply because with 9 dimensions, when one is curled up, two sets of the strings are identical, thus allowing for them to be eliminated.

The negative is that T duality still includes a perturbative duality. So, there was still another level needed to leap over the perturbative and non-perturbative regions. Researchers then made another breakthrough, called the S-duality. It was this duality that provided the necessary connection between non-perturbative and perturbative regions. But this breakthrough, like many in Quantum Physics, led to still another breakthrough, which is known as the U-duality. This breakthrough was even more powerful, but then duality made another leap.

Researchers showed that duality could be solved for the non-perturbative region in at least four different dimensional supersymmetric theories. Two researchers even found a duality between the 10-dimensional type IIA and 11-dimension strings, thus revealing supergravity. Thus, what was once a region that seemed off-limits was now open to further research. It also appeared that this region was governed by an 11-dimensional supergravity theory, but with a dimension curled up.

How does all this work when it comes to string theory? Simply put, scientists and researchers began to realize that string theory might not belong with the 10 dimensions, but the best explanation was 11 dimensions. It also means that this theory, at its fundamental core, is not string theory. Instead, this was a theory that could serve multiple purposes, such as being reduced down to 11 dimensional (supergravity) or 10-dimensional string theory, in addition to the p-brane theory. Still, nothing comes easy with Quantum Physics. Every breakthrough has those that look at it with a critical eye. So, what have these critics found that causes them to doubt this latest breakthrough?

Critics claim that the mathematical developments discussed here do not answer one fundamental question: how can you create an experiment to test it and either confirm or deny the validity of the theory. At this point, string theory can be described as the theory of Creation, a picture of beautiful symmetries in all their amazing glory. So, the only way to truly test it would be for the Big Bang to occur all over again. Many scientists argue that this is impossible to do, so there might be no other way to test it.

One scientist, a Nobel Laureate Sheldon Glashow, is critical of superstring theory, comparing it to other political plans that are untestable, drain resources, and siphons some of the best scientific minds to work on a fruitless task.

Those who support the superstring theory have suggested that their critics are missing the real point of this theory. Suppose this theory is solved using pure mathematics and thus solved non-perturbatively. In that case, the theory reduces downward, especially at low energies to the type of theory that functions with ordinary protons, atoms, and other molecules. At this stage, there are plenty of experimental data available.

Thus, the problem is that we haven't yet figured out how to write down this M-theory, solve it and finally understand what all of this means. For those who support string theory, particularly the M-theory, believe it is a question of brainpower. Also, by completely solving the theory, researchers and scientists can extract their low energy spectrum.

But now the question is, where should we be applying the brainpower to solve this theory for good, thus ending all the debates? Several areas could be points of attack. The most direct would be to try to derive using the Standard Model with all of its different particle interactions. However, researchers point to the fact that this theory, while successful, is one of the ugliest in terms of its bizarre and varied collection of quarks, electrons, and many other particle types. Yet it still might be possible by curling up at least 6 of the ten dimensions, creating a 4-dimensional theory that would have some similarity to the Standard Model. After this is done, you could then use duality and M-theory to begin a non-perturbative region probe. If the symmetries break properly or correctly, they should give us the correct masses of all the different particles that can be found in the Standard Model. Others argue that this isn't the way to get to the bottom of this mystery. Instead, they believe that solving string theory will involve understanding the basic underlying principles that are behind the theory. One example is how Einstein first came up with the theory of general relativity. When he realized that someone in an elevator that is falling would not feel gravity, he was able to extract from this line of thought the Equivalence Principle. The simple statement that physics laws are locally indistinguishable in any gravitating or accelerating frame allowed Einstein to introduce a new direction to physics, otherwise known as coordinate transformations.

Like any other science field, one thing inevitably leads to another. In our example, the next step was the action principle that was behind general relativity, becoming the most compelling theory of gravity. With string theory, we need to find the parallel to the Equivalence Principle, but for this theory instead.

In many ways, string theory has been developed in the wrong direction, starting in Quantum Theory before the other aspects of action, symmetry and principle were developed. A theory for the next century has fallen into the laps of those who might not be ready for it. If all that is the case, will this theory ever be written and thus a way to test it developed? Often, the answer is just another breakthrough away. For now, let's see what some other researchers have been able to do with the current knowledge and some of the potential theories they have spawned.

One researcher named Vafa added a twist to this story by introducing another mega-theory that was called F-theory. This new mega theory was based on a 12-dimensional explanation of self-duality within the IIb string. However, this theory has some problems, the first being that it has two-time coordinates, thus violating its 12-dimensional relativity. While it might not be the perfect solution, it begs the question of what the final theory might include, either 10, 11, or 12 dimensions. But what if there is, in fact, no fixed dimension. Various scientists believe that the final version could be independent of dimensionality within the context of space and time. Thus, dimensions only come into play when you try to solve the equation.

At this point, a full theory has yet to materialize. Still, many scientists and researchers believe there is more to discover. Even with a discovery or breakthrough every decade, this could be a theory that still takes a long time to complete and solve, thus finally understanding both the strings and how they (and everything else!) were created.

Chapter 19: Black Holes

Black Holes and the Mystery of Quantum Gravity

We talked about Einstein's theory of general relativity in the previous chapters. However, Einstein's theory is technically irreconcilable with quantum physics, for several reasons that we've mentioned in this book. According to Einstein, black holes are regions of ultra-intense gravity within spacetime that are so powerful, they pull even light inside.

The idea of a black hole was first proposed by a clergyman named John Mitchell in 1784. His ideas were dismissed for the most part because, shortly afterward, the light was "discovered" to be a wave instead of a particle (at the time). Therefore, scientists were unsure if gravity would be able to act on light "waves" instead of "particles" and thus largely forgot about the concept. That is until Einstein brought them back in 1915 with his general theory of relativity. From there, exactly how black holes are formed, how they work, and how they influence space and time have been hotly debated by many physicists. Scientists do believe the time is so distorted in a black hole that, to an outside observer, time would appear to stop inside of it. However, if a person were to fall into a black hole, time would proceed for them normally (resulting in a rather gruesome death).

Black holes can be created in a few different ways. One way would be by violating Planck's law. This has to do with the Planck constant, which you learned about before. Planck's constant also defines several measurements within quantum physics, and these measurements denote certain limits of measurement that can be achieved within the quantum universe. For example, if you tried to measure a particle's position with a laser at a greater accuracy than one Planck length (1.6×10^{-35} meters, which is very, *very* small), the laser's power would end up creating a very small black hole. Ironically, the black hole created would be exactly the size of one Planck length. Since time and space are intertwined, a black hole can also be created when trying to measure a length of time less than one unit of Planck time (10^{-43} seconds, which is very short).

Black holes are also, more traditionally, created by the collapse of very large stars. At the end of a star's life cycle, when the fuel inside the star has all run out, the mass of the star itself causes it to collapse, and all the matter inside of the star gets sucked in. Sometimes, this can create smaller, dimmer stars or create quasars (a type of ultra-bright, fast spinning, a very small star that also contains a black hole). Remember Heisenberg's uncertainty principle? Theoretically, this also can apply to the creation of black holes. If you'll remember, according to the uncertainty principle, it's impossible to determine a quantum particle's momentum and position with high precision at the same time. If you attempt to measure one or the other more precisely, the other measurement becomes less and less precise. If you get precise enough – say, down to one Planck length or less – the other value becomes so large that the particle is mathematically capable of turning into a black hole. This doesn't mean the particle is becoming a black hole, but more of a math problem illustrating the difficulty of measuring quantum particles. This is called an absurdity or a nonsense result – according to physicists, this means that something is missing from the math. They're missing a formula, a variable, or something else. This is the mystery of quantum gravity – even today, we're still missing something in the math required to make gravity (and, by extension, black holes) make sense.

These issues are propagated by the fact that there is no way to measure gravity on a coordinate plane. Since gravity exists within four dimensions, it can't be figured out using the two-dimensional math that scientists have traditionally used to figure out other quantum theories. Additionally, if you try to apply four-dimensional gravity to these two-dimensional theories, you receive anomalous answers – the math just doesn't make sense. Essentially, gravity *is* spacetime. As a result, you can't anchor any math within spacetime to help figure it out.

At its core, the reason why this math doesn't work is that, as we mentioned at the beginning of the chapter, Einstein's general theory of relativity is incompatible with quantum mechanics on a basic level.

The general theory of relativity works at a level that we can see, but at the quantum level (i.e., below the Planck scale), it unravels. This brings us full circle because that's why quantum mechanics was created in the first place. However, quantum mechanics has yet to account for the anomaly of gravity. For now, physicists continue to assume that we simply haven't found the right theory to solve it yet. As we mentioned in a previous chapter, many scientists believe string theory may be one of the candidates to do this.

As we mentioned a little earlier in this chapter, according to Einstein's theory of general relativity, black holes are essentially anomalies in spacetime that are holes with infinite mass. Black holes (and everything in the universe that experiences gravity, to a lesser extent) experience something called gravitational time dilation. This means that the closer you are to the black hole center, the slower time moves to an observer outside of the black hole. However, strangely enough, time will proceed at a normal pace to those closest to the black hole. This means that, if a pair of twins were to stand by a black hole – one twin very close to the black hole, and the other very far away – when the twins both came back to earth, the twin who was further from the black hole will have aged more than the one who was close to the black hole. Crazy, right? This also happens, to a lesser extent, to the satellites orbiting the earth, as well as the international space station (the ISS). Clocks that orbit Earth from space will slowly move ahead of clocks on earth.

Even stranger – continuing with the twin metaphor – if one of the twins were to fall inside of the black hole, and the other twin was to watch, time would appear to completely stop on the twin inside the black hole as the twin crossed over the edge. The twin watching outside the black hole would just see their other twin stop, freeze totally in time, then never move again, no matter how long the other twin waited for them. No one knows whether it would be possible to escape a black hole after this had happened (according to all known laws of physics, any normal person going into a black hole would be crushed by the incredibly strong gravitational force long before they made it to the center, of course) if it were possible to survive the experience. One theory is the white hole theory, which says that for every black hole, there is a matching white hole somewhere else in the galaxy that spews out all of the matter consumed by the black hole. If the person who fell into the black hole survived, then according to this theory, the black hole would work as a teleportation device. However, although we have documented cases of proposed white holes in the universe (the white hole GRB 060614 was found in 2006), there's no way to prove that these holes are in any way connected to black holes.

Some scientists have proposed that the Big Bang that created the universe was a white hole. The same paper proposes that white holes should be spontaneous, limited occurrences, rather than long-lasting singularities like black holes.

Conclusion

Thank you for making it through to the end of this book. Quantum physics is an endlessly fascinating subject and one that we think everyone should learn about in some capacity. After all, as we discover more and more about quantum physics, more of the future is open to us – things like teleportation, supercomputing, and ultra-fast space travel all feel like they're just around the corner if we can finally unravel how quantum mechanics works. Of course, we're not all scientists, and we can't all conduct experiments to figure out the next big discovery, but we can all do our part just by learning a little bit about how the universe around us works.

There are so many things going on in the universe around us that we can't explain. However, there's beauty in this un-knowing, and there's an indelible fascination in the discovery of the world around us. How boring would it be to already know exactly how everything around us worked? If we knew that, we would be able to predict everything, from whether there was alien life in the far reaches of the universe, to when our planet and solar system would die. While some might argue that it might be better to know, the universe's vastness allows for infinite new things to discover, crazy new laws and theorems to figure out, and new phenomena just over the event horizon. This information has assisted in the understanding of how stars are born, what matter and force do when they interact with each other on a particle level and also in larger masses.

The next step is to incorporate the knowledge you've gained here into your everyday life. Whether you use your new knowledge simply to brag, to help others, to do your research, to deepen your understanding of our world, or to become the next Albert Einstein, we're sure it will enrich your life in new and exciting ways. Please never stop reading, and more importantly, never stop learning.

Manufactured by Amazon.ca
Bolton, ON